Version 2

Viewing Life Mathematically
+ Integrated Review

Guided Notebook

Editors:
Robin Hendrix,
S. Rebecca Johnson,
Barbara Miller,
Amber Widmer

Copy Editors:
Mary Janelle Cady,
Rebecca Lebeaux

Creative Director:
Tee Jay Zajac

Designers:
Natalie Ezabele,
Trudy Gove,
Patrick Thompson

Cover Design:
Patrick Thompson

**Composition and
Answer Key Assistance:**
QSI (Pvt.) Ltd.

VP Research & Development: Marcel Prevuznak

Director of Content: Kara Roché

A division of Quant Systems, Inc.

546 Long Point Road, Mount Pleasant, SC 29464

Printed in the United States of America 🇺🇸

10 9 8 7 6 5 4 3 2

ISBN: 978-1-64277-099-5

Table of Contents

Strategies for Academic Success

CHAPTER 1.R

Review Concepts
for *Critical Thinking and Problem Solving*

CHAPTER 2.R

Review Concepts
for *Set Theory*

CHAPTER 12.R

Review Concepts
for *Sports*

CHAPTER 13.R

Review Concepts
for *Graph Theory*

CHAPTER 14.R

Review Concepts
for *Number Theory*

Math@Work

Math Knowledge Required for Math@Work Career Explorations

The following table summarizes the math knowledge required for each Math@Work career exploration. Use this table to determine when you are ready to explore each career.

Math@Work Career	Whole Numbers	Fractions	Integers	Decimal Numbers	Averages	Percents	Simple Interest	Ratios	Proportions	Geometry	Statistics	Graphing	Linear Equations	Systems of Equations	Mixture Problems	Scientific Notation	Greatest Common Factor	Rational Expressions	Radicals
Basic Inventory Management	✓																		
Hospitality Management	✓	✓			✓														
Bookkeeper				✓															
Pediatric Nurse				✓				✓	✓										
Architecture				✓						✓									
Statistician: Quality Control				✓							✓	✓							
Dental Assistant				✓	✓														
Financial Advisor				✓	✓	✓							✓						
Market Research Analyst				✓									✓						
Chemistry				✓										✓	✓				
Astronomy				✓												✓			
Math Education			✓														✓		
Physics		✓																✓	
Forensic Scientist				✓	✓														✓
Other Careers in Mathematics																			

⊗ Support

If you have questions or comments we can be contacted as follows:

24/7 Chat: chat.hawkeslearning.com

Phone: (843) 571-2825

E-mail: support@hawkeslearning.com

Web: hawkeslearning.com

Our support hours are 8:00 a.m. to 10:00 p.m. (ET), Monday through Friday.

How to Use the Guided Notebook

There are a variety of elements in this Guided Notebook that will help you on your way to mastering each topic. Here is a rundown of how to use the elements as you work through this notebook.

Fill-in-the-Blanks

1. When there is an incomplete sentence, you will need to write in the _____**missing**_____ word(s).

   The _____**missing**_____ words can be found by reading through the Learn screens.

Boxed Content

Definitions and **procedures** are highlighted within a box, like the ones shown here. The missing content will vary from box to box. Sometimes an entire definition is missing and sometimes only part of a sentence is missing. Here are two examples of the box variations.

Definition

First term to define: _____*Write the definition here.*_____

Second term to define: _____*If there is another term, define it the same way*_____
*as above.*_____

                                                                    DEFINITION

Terms Related to Probability

_____*Outcome*_____	An individual result of an experiment.
_____*Sample Space*_____	The set of all possible outcomes of an experiment.
_____*Event*_____	Some (or all) of the outcomes from the sample space.

                                                                    DEFINITION

Properties and **Procedure** boxes are completed in a similar way:

Commutative Property of Multiplication

The order of the numbers in multiplication can be <u>**reversed without changing the product.**</u>

For example, <u>**3 · 4 = 12   and   4 · 3 = 12.**</u>

PROPERTIES

Subtracting Whole Numbers

1. Write the numbers <u>**vertically**</u> so that the <u>**place values are lined up in columns.**</u>

2. Subtract only the <u>**digits with the same place value.**</u>

3. Check by <u>**adding the difference to the subtrahend.**</u> The sum must be <u>**the minuend.**</u>

PROCEDURE

▶ Watch and Work

For each Watch and Work, you will need to watch the corresponding video in Learn mode and follow along while completing the example in the space provided.

Example 5 Multiplying Whole Numbers

Multiply: $12 \cdot 35$

Solution

The standard form of multiplication is used here to find the product $12 \cdot 35$.

$$
\begin{array}{r}
\overset{1}{1}2 \\
\times\ 35 \\
\hline
60 \\
360 \\
\hline
420
\end{array}
\qquad
\begin{array}{l}
\\
\\
12 \cdot 5 = 60 \\
12 \cdot 30 = 360 \\
\text{Product}
\end{array}
$$

✏ Now You Try It!

After working along with the example video, work through a similar exercise on your own in the space provided.

Example A Multiplying Whole Numbers

Multiply: 25
         $\times\,42$
         1050

1.1 Exercises

Each section has exercises to offer additional practice problems to help reinforce topics that have been covered. The exercises include Concept Check, Practice, Application, and Writing & Thinking questions. The odd answers can be found in the Answer Key at the back of the book.

Concept Check

True/False. Determine whether each statement is true or false. If a statement is false, explain how it can be changed so the statement will be true. (**Note:** There may be more than one acceptable change.)

1. When the given statement is true, you write "True" for the answer.

   True

Practice

For each set of data, find **a.** the mean, **b.** the median, **c.** the mode (if any), and **d.** the range.

2. *Presidents:* The ages of the first five US presidents of the 20th century on the date of their inaugurations were as follows. (The presidents were Roosevelt, Taft, Wilson, Harding, and Coolidge.)

   42,   51,   56,   55,   51

   a. 51   b. 51   c. 51   d. 14

Applications

Solve.

3. *Grades:* Suppose that you have taken four exams and have one more chemistry exam to take. Each exam has a maximum of 100 points and you must average between 75 and 82 points to receive a passing grade of C. If you have scores of 85, 60, 73, and 76 on the first four exams, what is the minimum score you can make on the fifth exam and receive a grade of C?

   81

Writing & Thinking

4. State how to determine the median of a set of data.

   The first step to finding the median is always to arrange the data in order. Once the data is in order, the median is the number in the middle. If there is an even number of items, average the two middle numbers to find the median.

Strategies for
Academic Success

Strategies for Academic Success 🎓

How to Read a Math Textbook

Reading a textbook is very different than reading a book for fun. You have to concentrate more on what you are reading because you will likely be tested on the content. Reading a math textbook requires a different approach than reading literature or history textbooks because the math textbook contains a lot of symbols and formulas in addition to words. Here are some tips to help you successfully read a math textbook.

Don't Skim 📕

When reading math textbooks, look at everything: titles, learning objectives, definitions, formulas, text in the margins, and any text that is highlighted, outlined, or in bold. Also pay close attention to any tables, figures, charts, and graphs.

Minimize Distractions

Reading a math textbook requires much more concentration than a novel by your favorite author, so pick a study environment with few distractions and a time when you are most attentive.

🚩 Start at the Beginning

Don't start in the middle of an assigned section. Math tends to build on previously learned concepts and you may miss an important concept or formula that is crucial to understanding the rest of the material in the section.

Highlight and Annotate

Put your book to good use and don't be afraid to add comments and highlighting. If you don't understand something in the text, reread it a couple of times. If it is still not clear, note the text with a question mark or some other notation so you can ask your instructor about it.

Go through Each Step of Each Example 📊

Make sure you understand each step of an example. If you don't understand something, mark it so you can ask about it in class. Sometimes math textbooks leave out intermediate steps to save space. Try working through the examples on your own, filling in any missing steps.

Take Notes < *This is important!*

Write down important definitions, symbols or notation, properties, formulas, theorems, and procedures. Review these daily as you do your homework and before taking quizzes and tests. Practice rewriting definitions in your own words so you understand them better.

Notes 9-25-17:

- *The opposite of a negative integer is a positive integer.*

- *To add two integers with the same signs add their absolute values and use their common sign*

💻 Use Available Resources

Many textbooks have companion websites to help you understand the content. These resources may contain videos that help explain more complex steps or concepts. Try searching the internet for additional explanations of topics you don't understand.

Read the Material Before Class

Try to read the material from your book before the instructor lectures on it. After the lecture, reread the section again to help you retain the information as you look over your class notes.

Understand the Mathematical Definitions + × =

Many terms used in everyday English have a different meaning when used in mathematics. Some examples include equivalent, similar, average, median, and product. Two equations can be equivalent to one another without being equal. An average can be computed mathematically in several ways. It is important to note these differences in meaning in your notebook along with important definitions and formulas.

Try Reading the Material Aloud

Reading aloud makes you focus on every word in the sentence. Leaving out a word in a sentence or math problem could give it a totally different meaning, so be sure to read the text carefully and reread, if necessary.

Questions

1. Explain how taking notes can help you understand new concepts and skills while reading a math textbook.

2. Think of two more tips for reading a math textbook.

Strategies for Academic Success 🎓

Tips for Success in a Math Course

Read Your Textbook/Workbook

One of the most important skills when taking a math class is knowing how to read a math textbook. Reading a section before class and then reading it again afterwards is an important strategy for success in a math course. If you don't have time to read the entire assigned section, you can get an overview by reading the introduction or summary and looking at section objectives, headings, and vocabulary terms.

Take Notes 📝

Take notes in class using a method that works for you. There are many different note-taking strategies, such as the Cornell Method and Concept Mapping. You can try researching these and other methods to see if they might work better than your current note-taking system.

Review

While the information is fresh in your mind, read through your notes as soon as possible after class to make sure they are readable, write down any questions you have, and fill in any gaps. Mark any information that is incomplete so that you can get it from the textbook or your instructor later.

📁 Stay Organized

As you review your notes each day, be sure to label them using categories such as definition, theorem, formula, example, and procedure. Try highlighting each category with a different colored highlighter.

Use Study Aids

Use note cards to help you remember definitions, theorems, formulas, or procedures. Use the front of the card for the vocabulary term, theorem name, formula name, or procedure description. Write the definition, the theorem, the formula, or the procedure on the back of the card, along with a description in your own words.

Practice, Practice, Practice!

Math is like playing a sport. You can't improve your basketball skills if you don't practice—the same is true of math. Math can't be learned by only watching your instructor work through problems; you have to be actively involved in doing the math yourself. Work through the examples in the book, do some practice exercises at the end of the section or chapter, and keep up with homework assignments on a daily basis.

🖩 Do Your Homework

When doing homework, always allow plenty of time to finish it before it is due. Check your answers when possible to make sure they are correct. With word or application problems, always review your answer to see if it appears reasonable. Use the estimation techniques that you have learned to determine if your answer makes sense.

Understand, Don't Memorize

Don't try to memorize formulas or theorems without understanding them. Try describing or explaining them in your own words or look for patterns in formulas so you don't have to memorize them. For example, you don't need to memorize every perimeter formula if you understand that perimeter is equal to the sum of the lengths of the sides of the figure.

Study

Plan to study two to three hours outside of class for every hour spent in class. If math is your most difficult subject, then study while you are alert and fresh. Pick a study time when you will have the least interruptions or distractions so that you can concentrate.

🕐 Manage Your Time

Don't spend more than 10 to 15 minutes working on a single problem. If you can't figure out the answer, put it aside and work on another one. You may learn something from the next problem that will help you with the one you couldn't do. Mark the problems that you skip so that you can ask your instructor about it during the next class. It may also help to work a similar, but perhaps easier, problem.

Questions

1. Based on your schedule, what are the best times and places for you to study for this class?

2. Describe your method of taking notes. List two ways to improve your method.

Strategies for Academic Success 🎓

Tips for Improving Math Test Scores

Preparing for a Math Test

- Avoid cramming right before the test and don't wait until the night before to study. Review your notes and note cards every day in preparation for quizzes and tests.

- If the textbook has a chapter review or practice test after each chapter, work through the problems as practice for the test.

- If the textbook has accompanying software with review problems or practice tests, use it for review.

- Review and rework homework problems, especially the ones that you found difficult.

- If you are having trouble understanding certain concepts or solving any types of problems, schedule a meeting with your instructor or arrange for a tutoring session (if your college offers a tutoring service) well in advance of the next test.

Test-Taking Strategies

- Scan the test as soon as you get it to determine the number of questions, their levels of difficulty, and their point values so you can adequately gauge how much time you will have to spend on each question.

- Start with the questions that seem easiest or that you know how to work immediately. If there are problems with large point values, work them next since they count for a larger portion of your grade.

- Show all steps in your math work. This will make it quicker to check your answers later once you are finished since you will not have to work through all the steps again.

- If you are having difficulty remembering how to work a problem, skip it and come back to it later so that you don't spend all of your time on one problem.

After the Test

- The material learned in most math courses is cumulative, which means any concepts you miss on each test may be needed to understand concepts in future chapters. That's why it is extremely important to review your returned tests and correct any misunderstandings that may hinder your performance on future tests.

- Be sure to correct any work you did wrong on the test so that you know the correct way to do the problem in the future. If you are not sure what you did wrong, get help from a peer who scored well on the test or schedule time with your instructor to go over the test.

- Analyze the test questions to determine if the majority came from your class notes, homework problems, or the textbook. This will give you a better idea of how to spend your time studying for the next test.

- Analyze the errors you made on the test. Were they careless mistakes? Did you run out of time? Did you not understand the material well enough? Were you unsure of which method to use?

- Based on your analysis, determine what you should do differently before the next test and where you should focus your time.

Questions

1. Determine the resources that are available to you to help you prepare for tests, such as instructor office hours, tutoring center hours, and study groups.

2. Discuss two additional test taking strategies.

Strategies for Academic Success 🎓

Practice, Patience, and Persistence!

Have you ever heard the phrase "practice makes perfect"? This saying applies to many things in life. You won't become a concert pianist without many hours of practice. You won't become an NBA basketball star by sitting around and watching basketball on TV. The saying even applies to riding a bike. You can watch all of the videos and read all of the books on riding a bike, but you won't learn how to ride a bike without actually getting on the bike and trying to do it yourself. The same idea applies to math. Math is not a spectator sport.

Math is not learned by sleeping with your math book under your pillow at night and hoping for osmosis (a scientific term implying that math knowledge would move from a place of higher concentration—the math book—to a place of lower concentration—your brain). You also don't learn math by watching your professor do hundreds of math problems while you sit and watch. Math is learned by doing. Not just by doing one or two problems, but by doing many problems. Math is just like a sport in this sense. You become good at it by doing it, not by watching others do it. You can also think of learning math like learning to dance. A famous ballerina doesn't take a dance class or two and then end up dancing the lead in The Nutcracker. It takes years of practice, patience, and persistence to get that part.

Now, we aren't suggesting that you dedicate your life to doing math, but at this point in your education, you've already spent quite a few years studying the subject. You will continue to do math throughout college—and your life. To be able to financially support yourself and your family, you will have to find a job, earn a salary, and invest your money—all of which require some ability to do math. You may not think so right now, but math is one of the more useful subjects you will study.

It's important not only to practice math when taking a math course, but also to be patient and not expect immediate success. Just like a ballerina or NBA basketball star, who didn't become exceptional athletes overnight, it will take some time and patience to develop your math skills. Sure, you will make some mistakes along the way, but learn from those mistakes and move on.

Practice, patience, and persistence are especially important when working through applications or word problems. Most students don't like word problems and, therefore, avoid them. You won't become good at working word problems unless you practice them over and over again. You'll need to be patient when working through word problems in math since they will require more time to work than typical math skills exercises. The process of solving word problems is not a quick one and will take patience and persistence on your part to be successful.

Just as you work your body through physical exercise, you have to work your brain through mental exercise. Math is an excellent subject to provide the mental exercise needed to stimulate your brain. Your brain is flexible and it continues to grow throughout your life span—but only if provided the right stimuli. Studying mathematics and persistently working through tough math problems is one way to promote increased brain function. So, when doing mathematics, remember the 3 P's—Practice, Patience, and Persistence—and the positive effects they will have on your brain!

Questions

1. What is another area (not mentioned here) that requires practice, patience, and persistence to master? Can you think of anything you could master without practice?

2. Can you think of an example in your study of math where practice, patience, and persistence have helped you improve?

Strategies for Academic Success 🎓

Note Taking

Taking notes in class is an important step in understanding new material. While there are several methods for taking notes, every note-taking method can benefit from these general tips.

General Tips

- Write the date and the course name at the top of each page.
- Write the notes in your own words and paraphrase.
- Use abbreviations, such as ft for foot, # for number, def for definition, and RHS for right-hand side.
- Copy all figures or examples that are presented during the lecture.
- Review and rewrite your notes after class. Do this on the same day, if possible.

There are many different methods of note taking and it's always good to explore new methods. A good time to try out new note-taking methods is when you rewrite your class notes. Be sure to try each new method a few times before deciding which works best for you. Presented here are three note-taking methods you can try out. You may even find that a blend of several methods works best for you.

Note-Taking Methods

Outline

An outline consists of several topic headings, each followed by a series of indented bullet points that include subtopics, definitions, examples, and other details.

> Example:
>
> **1.** Ratio
>   **a.** Comparison of two quantities by division.
>   **b.** Ratio of a to b
>   **i.** $\dfrac{a}{b}$
>   **ii.** $a : b$
>   **iii.** a to b
>   **c.** Can be reduced
>   **d.** Common units can cancel

Split Page

The split page method divides the page vertically into two columns with the left column narrower than the right column. Main topics go in the left column and detailed comments go in the right column. The bottom of the page is reserved for a short summary of the material covered.

> Example:
>
> | Keywords: | Notes: |
> |-----------|--------|
> | Ratios | 1. Comparison of two quantities by division
2. $\dfrac{a}{b}$, $a : b$, a to b
3. Can reduce
4. Common units can cancel |
>
> Summary: Ratios are used to compare quantities and units can cancel.

Mapping

The mapping method is the most visual of the three methods. One common way to create a mapping is to write the main idea or topic in the center and draw lines, from the main idea to smaller ideas or subtopics. Additional branches can be created from the subtopics until all of the key ideas and definitions are included. Using a different color for subtopic can help visually organize the topics.

> Example:
>
>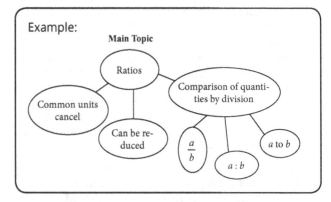

Questions

1. Find two other note taking methods and describe them.

2. Write five additional abbreviations that you could use while taking notes.

Strategies for Academic Success 🎓

Do I Need a Math Tutor?

If you do not understand the material being presented in class, if you are struggling with completing homework assignments, or if you are doing poorly on tests, then you may need to consider getting a tutor. In college, everyone needs help at some point in time. What's important is to recognize that you need help before it's too late and you end up having to retake the class.

Alternatives to Tutoring

Before getting a tutor, you might consider setting up a meeting with your instructor during their office hours to get help. Unfortunately, you may find that your instructor's office hours don't coincide with your schedule or don't provide enough time for one-on-one help.

Another alternative is to put together a study group of classmates from your math class. Working in groups and explaining your work to others can be very beneficial to your understanding of mathematics. Study groups work best if there are three to six members. Having too many people in a study group may make it difficult to schedule a time for all group members to meet. A large study group may also increase distractions. If you have too few people and those that attend are just as lost as you, then you aren't going to be helpful to each other.

Where to Find a Tutor

Many schools have both group and individual tutoring available. In most cases, the cost of this tutoring is included in tuition costs. If your college offers tutoring through a learning lab or tutoring center, then you should take advantage of it. You may need to complete an application to be considered for tutoring, so be sure to get the necessary paperwork at the start of each semester to increase your chances of getting a tutoring time that works well with your schedule. This is especially important if you know that you struggle with math or haven't taken any math classes in a while.

If you find that you need more help than the tutoring center can provide, or your school doesn't offer tutoring, you can hire a private tutor. The hourly cost to hire a private tutor varies significantly depending on the area you live in along with the education and experience level of the tutor. You might be able to find a tutor by asking your instructor for references or by asking friends who have taken higher-level math classes than you have. You can also try researching the internet for local reputable tutoring organizations in your area.

What to Look for in a Tutor

Whether you obtain a tutor through your college or hire a personal tutor, look for someone who has experience, educational qualifications, and who is friendly and easy to work with. If you find that the tutor's personality or learning style isn't similar to yours, then you should look for a different tutor that matches your style. It may take some effort to find a tutor who works well with you.

How to Prepare for a Tutoring Session

To get the most out of your tutoring session, come prepared by bringing your text, class notes, and any homework or questions you need help with. If you know ahead of time what you will be working on, communicate this to the tutor so they can also come prepared. You should attempt the homework prior to the session and write notes or questions for the tutor. Do not use the tutor to do your homework for you. The tutor will explain to you how to do the work and let you work some problems on your own while he or she observes. Ask the tutor to explain the steps aloud while working through a problem. Be sure to do the same so that the tutor can correct any mistakes in your reasoning. Take notes during your tutoring session and ask the tutor if he or she has any additional resources such as websites, videos, or handouts that may help you.

Questions

1. It's important to find a tutor whose learning style is similar to yours. What are some ways that learning styles can be different?

2. What sort of tutoring services does your school offer?

Strategies for Academic Success 🎓

Tips for Improving Your Memory

Experts believe that there are three ways that we store memories: first in the sensory stage, then in short term memory, and finally in long term memory.[1] Because we can't retain all the information that bombards us daily, the different stages of memory act as a filter. Your sensory memory lasts only a fraction of a second and holds your perception of a visual image, a sound, or a touch. The sensation then moves to your short term memory, which has the limited capacity to hold about seven items for no more than 20 to 30 seconds at a time. Important information is gradually transferred to long term memory. The more the information is repeated or used, the greater the chance that it will end up in long term memory. Unlike sensory and short term memory, long term memory can store unlimited amounts of information indefinitely. Here are some tips to improve your chances of moving important information to long-term memory.

Be attentive and focused on the information.

Study in a location that is free of distractions and avoid watching TV or listening to music with lyrics while studying.

Recite information aloud.

Ask yourself questions about the material to see if you can recall important facts and details. Pretend you are teaching or explaining the material to someone else. This will help you put the information into your own words.

Associate the information with something you already know.

Think about how you can make the information personally meaningful—how does it relate to your life, your experiences, and your current knowledge? If you can link new information to memories already stored, you create "mental hooks" that help you recall the information. For example, when trying to remember the formula for slope using rise and run, remember that rise would come alphabetically before run, so rise will be in the numerator in the slope fraction and run will be in the denominator.

Use visual images like diagrams, charts, and pictures.

You can make your own pictures and diagrams to help you recall important definitions, theorems, or concepts.

Split larger pieces of information into smaller "chunks."

This is useful when remembering strings of numbers, such as social security numbers and telephone numbers. Instead of remembering a sequence of digits such as 555777213 you can break it into chunks such as 555 777 213.

Group long lists of information into categories that make sense.

For example, instead of remembering all the properties of real numbers individually, try grouping them into shorter lists by operation, such as addition and multiplication.

Use mnemonics or memory techniques to help remember important concepts and facts.

A mnemonic that is commonly used to remember the order of operations is "Please Excuse My Dear Aunt Sally," which uses the first letter of the words Parentheses, Exponents, Multiplication, Division, Addition, and Subtraction to help you remember the correct order to perform basic arithmetic calculations. To make the mnemonic more personal and possibly more memorable, make up one of your own.

Use acronyms to help remember important concepts or procedures.

An acronym is a type of mnemonic device which is a word made up by taking the first letter from each word that you want to remember and making a new word from the letters. For example, the word HOMES is often used to remember the five Great Lakes in North America where each letter in the word represents the first letter of one of the lakes: Huron, Ontario, Michigan, Erie, and Superior.

Questions

1. Create an original mnemonic or acronym for any math topic covered so far in this course.

2. Explain two ways you can incorporate these tips into your study routine.

---

1   Source: http://science.howstuffworks.com/life/inside-the-mind/human-brain/human-memory2.htm

Strategies for Academic Success 🎓

Overcoming Anxiety

People who are anxious about math are often just not good at taking math tests. If you understand the math you are learning but don't do well on math tests, you may be in the same situation. If there are other subject areas in which you also perform poorly on tests, then you may be experiencing test anxiety.

How to Reduce Math Anxiety

- Learn effective math study skills. Sit near the front of your class and take notes. Ask questions when you don't understand the material. Review your notes after class and read new material before it's covered in class. Keep up with your assignments and do a lot of practice problems.

- Don't accept negative self talk such as "I am not good at math" or "I just don't get it and never will." Maintain a positive attitude and set small math achievement goals to keep you positively moving toward bigger goals.

- Visualize yourself doing well in math, whether it's on a quiz or test, or passing a math class. Rehearse how you will feel and perform on an upcoming math test. It may also help to visualize how you will celebrate your success after doing well on the test.

- Form a math study group. Working with others may help you feel more relaxed about math in general and you may find that other people have the same fears.

- If you panic or freeze during a math test, try to work around the panic by finding something on the math test that you can do. Once you gain confidence, work through other problems you know how to do. Then, try completing the harder problems, knowing that you have a large part of the test completed already.

- If you have trouble remembering important concepts during tests, do what is called a "brain drain" and write down all the formulas and important facts that you have studied on your test or scratch paper as soon as you are given the test. Do this before you look at any questions on the test. Having this information available to you should help boost your confidence and reduce your anxiety. Doing practice brain drains while studying can help you remember the concepts when the test time comes.

How to Reduce Test Anxiety

- Be prepared. Knowing you have prepared well will make you more confident and less anxious.

- Get plenty of sleep the night before a big test and be sure to eat nutritious meals on the day of the test. It's helpful to exercise regularly and establish a set routine for test days. For example, your routine might include eating your favorite food, putting on your lucky shirt, and packing a special treat for after the test.

- Talk to your instructor about your anxiety. Your instructor may be able to make accommodations for you when taking tests that may make you feel more relaxed, such as extra time or a more calming testing place.

- Learn how to manage your anxiety by taking deep, slow breaths and thinking about places or people who make you happy and peaceful.

- When you receive a low score on a test, take time to analyze the reasons why you performed poorly. Did you prepare enough? Did you study the right material? Did you get enough rest the night before? Resolve to change those things that may have negatively affected your performance in the past before the next test.

- Learn effective test taking strategies. See the study skill on Tips for Improving Math Test Scores.

Questions

1. Describe your routine for test days. Think of two ways you can improve your routine to reduce stress and anxiety.

2. Research and describe the accommodations that your instructor or school can provide for test taking.

Strategies for Academic Success 🎓

Online Resources

With the invention of the internet, there are numerous resources available to students who need help with mathematics. Here are some quality online resources that we recommend.

HawkesTV

tv.hawkeslearning.com

If you are looking for instructional videos on a particular topic, then start with HawkesTV. There are hundreds of videos that can be found by looking under a particular math subject area such as introductory algebra, precalculus, or statistics. You can also find videos on study skills.

YouTube

www.youtube.com

You can also find math instructional videos on YouTube, but you have to search for videos by topic or key words. You may have to use various combinations of key words to find the particular topic you are looking for. Keep in mind that the quality of the videos varies considerably depending on who produces them.

Google Hangouts

hangouts.google.com

You can organize a virtual study group of up to 10 people using Google Hangouts. This is a terrific tool when schedules are hectic and it avoids everyone having to travel to a central location. In addition to video chat, the group members can share documents using Google Docs. This is a great tool for group projects!

Wolfram|Alpha

www.wolframalpha.com

Wolfram|Alpha is a computational knowledge engine developed by Wolfram Research that answers questions posed to it by computing the answer from "curated data." Typical search engines search all of the data on the Internet based on the key words given and then provide a list of documents or web pages that might contain relevant information. The data used by Wolfram|Alpha is said to be "curated" because someone has to verify its integrity before it can be added to the database, therefore ensuring that the data is of high quality. Users can submit questions and request calculations or graphs by typing their request into a text field. Wolfram|Alpha then computes the answers and related graphics from data gathered from both academic and commercial websites such as the CIA's World Factbook, the United States Geological Survey, financial data from Dow Jones, etc. Wolfram|Alpha uses the basic features of Mathematica, which is a computational toolkit designed earlier by Wolfram Research that includes computer algebra, symbol and number computation, graphics, and statistical capabilities.

Questions

1. Describe a situation where you think Wolfram|Alpha might be more helpful than YouTube, and vice versa.

2. What are some pros and cons to using Google Hangouts?

Strategies for Academic Success 🎓

Preparing for a Final Math Exam

Since math concepts build on one another, a final exam in math is not one you can study for in a night or even a day or two. To pull all the concepts together for the semester, you should plan to start one or two weeks ahead of time. Being comfortable with the material is key to going into the exam with confidence and lowering your anxiety.

Before You Start Preparing for the Exam

1. What is the date, time, and location of the exam? Check your syllabus for the final exam time and location. If it's not on your syllabus, your instructor should announce this information in class.

2. Is there a time limit on the exam? If you experience test anxiety on timed tests, be sure to speak to your professor about it and see if you can receive accommodations that will help reduce your anxiety, such as extended time or an alternate testing location.

3. Will you be able to use a formula sheet, calculator, and/or scrap paper on the exam? If you are not allowed to use a formula sheet, you should write down important formulas and memorize them. Most of the time, math professors will advise you of the formulas you need to know for an exam. If you cannot use a calculator on the exam, be sure to practice doing calculations by hand when you are preparing for the exam and go back and check them using the calculator.

A Week Before the Exam

1. Decide where to study for the exam and with whom. Make sure it's a comfortable study environment with few outside distractions. If you are studying with others, make sure the group is small and that the people in the group are motivated to study and do well on the exam. Plan to have snacks and water with you for energy and to avoid having to delay studying to go get something to eat or drink. Be sure and take small breaks every hour or two to keep focused and minimize frustration.

2. Organize your class notes and any flash cards with vocabulary, formulas, and theorems. If you haven't used flash cards for vocabulary, go back through your notes and highlight the vocabulary. Create a formula sheet to use on the exam, if the professor allows. If not, then you can use the formula sheet to memorize the formulas that will be on the exam.

3. Start studying for the exam. Studying a week before the exam gives you time to ask your instructor questions as you go over the material. Don't spend a lot of time reviewing material you already know. Go over the most difficult material or material that you don't understand so you can ask questions about it. Be sure to review old exams and work through any questions you missed.

3 Days Before the Exam

1. Make yourself a practice test consisting of the problem types. Don't necessarily put the questions in the order that the professor covered them in class.

2. Ask your instructor or classmates any questions that you have about the practice test so that you have time to go back and review the material you are having difficulty with.

The Night Before the Exam

1. Make sure you have all the supplies you will need to take the exam: formula sheet and calculator, if allowed, scratch paper, plain and colored pencils, highlighter, erasers, graph paper, extra batteries, etc.

2. If you won't be allowed to use your formula sheet, review it to make sure you know all the formulas. Right before going to bed, review your notes and study materials, but do not stay up all night to "cram."

3. Go to bed early and get a good night's sleep. You will do better if you are rested and alert.

The Day of the Exam

1. Get up with plenty of time to get to your exam without rushing. Eat a good breakfast and don't drink too much caffeine, which can make you anxious.

2. Review your notes, flash cards, and formula sheet again, if you have time.

3. Get to class early so you can be organized and mentally prepared.

Checklist for the Exam

Date of the Exam: _____   Time of the Exam: _____

Location of the Exam: _____

Items to bring to the exam:

___ calculator and extra batteries   ___ pencils

___ formula sheet   ___ eraser

___ scratch paper   ___ colored pencils or highlighter

___ graph paper   ___ ruler or straightedge

Notes or other things to remember for exam day:

During the Exam

1. Put your name at the top of your exam immediately. If you are not allowed to use a formula sheet, before you even look at the exam, do what is called a "brain drain" or "data dump." Recall as much of the information on your formula sheet as you possibly can and write it either on the scratch paper or in the exam margins if scratch paper is not allowed. You have now transferred over everything on your "mental cheat sheet" to the exam to help yourself as you work through the exam.

2. Read the directions carefully as you go through the exam and make sure you have answered the questions being asked. Also, check your solutions as you go. If you do any work on scratch paper, write down the number of the problem on the paper and highlight or circle your answer. This will save you time when you review the exam. The instructor may also give you partial credit for showing your work. (Don't forget to attach your scratch work to your exam when you turn it in.)

3. Skim the questions on the exam, marking the ones you know how to do immediately. These are the problems you will do first. Also note any questions that have a higher point value. You should try to work these next or be sure to leave yourself plenty of time to do them later.

4. If you get to a problem you don't know how to do, skip it and come back after you finish all the ones you know how to do. A problem you do later may jog your memory on how to do the problem you skipped.

5. For multiple choice questions, be sure to work the problem first before looking at the answer choices. If your answer is not one of the choices, then review your math work. You can also try starting with the answer choices and working backwards to see if any of them work in the problem. If this doesn't work, see if you can eliminate any of the answer choices and make an educated guess from the remaining ones. Mark the problem to come back to later when you review the exam.

6. Once you have an answer for all the problems, review the entire exam. Try working the problems differently and comparing the results or substituting the answers into the equation to verify they are correct. Do not worry about finishing early. You are in control of your own time—and your own success!

Questions

1. Does your syllabus provide any of the information needed for the checklist?

2. Are there any tips or suggestions mentioned here that you haven't thought of before?

Strategies for Academic Success 🎓

Managing Your Time Effectively

Have you ever made it to the end of a day and wondered where all of your time went? Sometimes it feels like there aren't enough hours in the day. Managing your time is important because you can never get that time back. Once it's gone, you have to rush and cram the work into your schedule. Not only will you start feeling stressed out, but you may also find yourself turning in late or incomplete work.

Here are three strategies for managing your time more effectively.

🕐 Time Budgets

Time budgets help you find the time you need to complete necessary projects and tasks. Just like a financial budget shows you how you spend your money, a time budget shows you how you spend your time. You can then identify "wasted" time that could be used more productively.

To begin budgeting your time, assess how much time each week you spend on different types of activities, like Sleep, Meals, Work, Class, Study, Extracurricular, Exercise, Personal, Other, etc.

- What are some activities you'd like to spend more time doing in the future?
- What are some activities you should spend less time doing in the future?

Based on your answers to the questions above, create a weekly time budget. One week contains only 168 hours. If you want to spend more time on a particular activity, you'll need to find that time somewhere. Use a planner to schedule specific blocks of time for study sessions, meals, travel times, and morning/evening routines. As a general rule, you should set aside at least two hours of study time for every one hour of class time. That means that a three-credit course would require at least six hours of outside work per week.

⚖️ Breaks

When you are working on an important project or studying for a big exam, you can feel tempted to go as long as possible without taking a break. While staying focused is important, working yourself until you're mentally drained will lower the quality of your work and force you to take even more time recovering.

Just like taking breaks helps your physical body recover, it will also help your brain re-energize and refocus. During study sessions, you should plan to take a break at least once an hour. Study and work breaks should usually last around five minutes. The longer the break, the harder it is to start working again. Some courses have a built-in break during the middle of the class period. Stand up and move around, even if you don't feel tired. Even this little bit of physical movement can help you think more clearly.

📧 Avoiding Multitasking

Multitasking is working on more than one task at a time. When you have several assignments that need to be completed, you may be tempted to save time by working on two or three of them at once. While this strategy might seem like a time-saver, you will probably end up using more time than if you had done each task individually. Not only will you have to switch your focus from one task to the next, but you will also make more mistakes that will need to be corrected later. Multitasking usually ends up wasting time instead of saving it.

Instead of trying to do two things at once, schedule yourself time to work on one task at a time. To-do lists can be helpful tools for keeping yourself focused on finishing one item before moving on to another. You'll do better work and save yourself time.

Questions

1. Are there any areas in your day that are taking up too much of your time, making it hard to devote enough time to more important things?

2. Can you think of a time when multitasking has resulted in lower quality outcome in your experience?

Review Concepts

for *Critical Thinking and Problem Solving*

1.R.1 Introduction to Whole Numbers

Whole Numbers

The **whole numbers** are _____ along with _____

Natural numbers = $\mathbb{N}$ = { _____ }

Whole numbers = $\mathbb{W}$ = { _____ }

DEFINITION

The Decimal System

The **decimal system** (or base ten system) is a place value system that depends on three things.

1. _____

2. the placement of _____

3. the value of _____

DEFINITION

1. In **expanded notation**, the values represented by each digit are _____

▶ Watch and Work

Watch the video for Example 3 in the software and follow along in the space provided.

Example 3 Writing Numbers in Expanded Notation

Write each number in expanded notation.

a. 954   b. 6507

Solution

✏ Now You Try It!

Use the space provided to work out the solution to the next example.

Example A Application: Finding the Mean

Write each number in expanded notation.

a. 463

b. 7300

Reading and Writing Whole Numbers

You should note the following four things when reading or writing whole numbers.

1. Digits are read in _____ (groups _____).

2. Commas are used to _____ if a number has _____

3. The word "and" does not _____
    _____

4. Hyphens are used to write words for _____
    _____

1.R.1 Exercises

Concept Check

True/False. Determine whether each statement is true or false. If a statement is false, explain how it can be changed so the statement will be true. (**Note:** There may be more than one acceptable change.)

1. In the number 21,057, the "1" represents 1000.

2. 56,317 can be written as $56{,}000 + 300 + 17$ in expanded notation.

3. 42,360 can be written as forty-two thousand, three hundred sixty.

4. The word "and" is not used when reading or writing whole numbers.

Practice

5. Name the place value of each nonzero digit in the following number: 24,608.

6. Write 1892 in expanded notation.

7. Write 683,100 in words.

Write each number in standard notation.

8. Write four hundred thousand, seven hundred thirty-six in standard notation.

Applications

9. *Lakes:* The largest lake in the United States is Lake Superior. It takes up an area of 82,103 square kilometers. Write 82,103 in words.

10. *Card Games:* The largest collection of Joker playing cards consists of eight thousand, five hundred twenty cards amassed by Tony De Santis after inheriting a two thousand piece collection from the magician Fernando Riccardi. Write eight thousand, five hundred twenty in standard form.

Writing & Thinking

11. How are natural numbers and whole numbers different and how are they the same?

12. When are hyphens used to write numbers in English words?

1.R.2 Rounding and Estimating with Whole Numbers

Rounding Numbers

To **round** a given number means _____

DEFINITION

Rounding Rule for Whole Numbers

1. Look at the single digit just to the right of the digit in the place of desired accuracy.

   a. **If this digit is less than 5**, leave the digit in the place of desired accuracy as it is, and _____

   _____

   b. **If this digit is 5 or greater**, increase the digit in the desired place of accuracy by _____

   _____ All digits to the left remain unchanged unless

   _____ Then the 9 is replaced by 0 and _____

   _____

PROCEDURE

To Estimate a Sum or Difference

1. Round each number to _____

2. Perform the _____

PROCEDURE

▶ Watch and Work

Watch the video for Example 4 in the software and follow along in the space provided.

Example 4 Estimating Sums of Whole Numbers

Estimate the sum; then find the actual sum.

$$
\begin{array}{r}
68 \\
925 \\
+487 \\
\hline
\end{array}
$$

Solution

✏️ Now You Try It!

Use the space provided to work out the solution to the next example.

Example A Estimating Sums of Whole Numbers

Estimate the sum; then find the actual sum.

$$
\begin{array}{r}
176 \\
84 \\
+\ 75 \\
\hline
\end{array}
$$

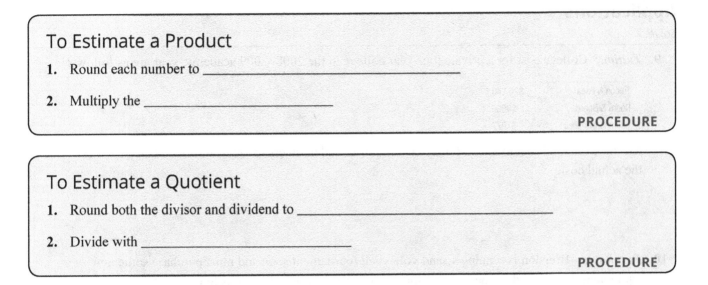

To Estimate a Product

1. Round each number to _____

2. Multiply the _____

PROCEDURE

To Estimate a Quotient

1. Round both the divisor and dividend to _____

2. Divide with _____

PROCEDURE

1.R.2 Exercises

Concept Check

True/False. Determine whether each statement is true or false. If a statement is false, explain how it can be changed so the statement will be true. (**Note:** There may be more than one acceptable change.)

1. Rounding means finding a number close to the given number, using a specified place of accuracy.

2. When rounded to the ten thousands place, 435,613 becomes 400,000.

3. To estimate the answer for a division problem, begin by rounding both the divisor and dividend.

4. If estimated, $4250 \div 51$ is $4000 \div 50 = 80$.

Practice

Estimate each answer; then find the actual answer.

5. $\begin{array}{r} 83 \\ 62 \\ + 78 \\ \hline \end{array}$

6. $\begin{array}{r} 63{,}504 \\ - 42{,}700 \\ \hline \end{array}$

7. $\begin{array}{r} 420 \\ \times 104 \\ \hline \end{array}$

8. $11\overline{)99}$

Applications

Solve.

9. *Tuition:* College cost for a private four-year college in the 2008–2009 academic year are as follows:

Tuition & Fees	$25,243
Room & Board	$8996
Books & Supplies	$1077

Estimate the total cost to attend for a year using rounded numbers to the nearest thousand. Then calculate the actual cost.

10. *Purchases:* Brendon is running a sand volleyball tournament soon and must purchase some new equipment. He needs three new nets, which cost $159 each. He also needs five new sets of boundary lines, which cost $86 each. Estimate the total cost of the new equipment. Then calculate the actual cost.

Writing & Thinking

11. In your own words, define estimation.

12. Compare and contrast rounding and estimating.

1.R.3 Exponents and Order of Operations

1. When looking at $3^5 = 243$, 3 is the _____, 5 is the _____, and 243

   is the _____. _____ are written slightly to the right and above

   the _____. The expression 3^5 is _____.

The Exponent 1

Any number raised to the first power _____

For example, _____

DEFINITION

The Exponent 0

Any nonzero number raised to the 0 power _____

For example, _____

Note: The expression 0^0 _____

DEFINITION

Rules for Order of Operations

1. Simplify within grouping symbols, such as _____
   (If there are more than one pair of grouping symbols, start with _____
   _____

2. Evaluate any _____

3. Moving from left to right, perform any _____
   _____

4. Moving from left to right, perform any _____
   _____

PROCEDURE

---

2. A well-known mnemonic device for remembering the rules for order of operations is the following.

Please	Excuse	My	Dear	Aunt	Sally
↓	↓	↓	↓	↓	↓
_____	_____	_____	_____	_____	_____

▶ Watch and Work

Watch the video for Example 6 in the software and follow along in the space provided.

Example 6 Using the Order of Operations with Whole Numbers

Simplify: $2 \cdot 3^2 + 18 \div 3^2$

Solution

✏ Now You Try It!

Use the space provided to work out the solution to the next example.

Example A Using the Order of Operations with Whole Numbers

Simplify: $6^2 \div 9 + 3 - 14 \div 7$

1.R.3 **Exercises**

Concept Check

True/False. Determine whether each statement is true or false. If a statement is false, explain how it can be changed so the statement will be true. (**Note:** There may be more than one acceptable change.)

1. Nine squared is equal to eighteen.

2. $2^7 = 128$

3. 7^0 is undefined.

4. According to the order of operations, multiplication is always performed before division.

Practice

For each exponential expression **a.** identify the base, **b.** identify the exponent, and **c.** evaluate the exponential expression.

5. 2^3

6. 4^0

Simplify.

7. $18 \div 2 - 1 - 3 \cdot 2$

8. $30 \div 2 - 11 + 2(5-1)^3$

Applications

Solve.

9. ***Card Games:*** Neville bought 15 boxes of trading cards. Each box has 10 packs of trading cards. Each pack of trading cards contains 20 cards. He adds 132 cards that he already owns to the newly purchased cards. Then, Neville evenly distributes all of the cards to 6 of his friends. How many trading cards would each person get?

   a. If you simplify the expression $15 \cdot 10 \cdot 20 + 132 \div 6$ using the order of operations, will you get the correct answer? If not, explain what is wrong with the expression.

   b. What is the answer? If necessary, write the corrected expression to get the correct results when following the order of operations.

10. ***Purchases:*** Robert is purchasing shirts for his weekend soccer team. The shirts he wants to buy are normally $25 each but are on sale for $10 off. His team has a total of 11 players. How much will he spend to buy the shirts?

   a. If you simplify the expression $\$25 - \$10 \cdot 11$ using the order of operations, will you get the correct answer? If not, explain what is wrong with the expression.

   b. What is the answer? If necessary, write the corrected expression to get the correct results when following the order of operations.

Writing & Thinking

11. Give one example where addition should be completed before multiplication.

1.R.4 Problem Solving with Whole Numbers

> ## Basic Strategy for Solving Word Problems
>
> 1. READ: _____
>
> 2. SET UP: Draw any type of figure or diagram that might be helpful and _____
>   _____
>
> 3. SOLVE: _____
>
> 4. CHECK: Check your work and _____
>
> **PROCEDURE**

> ## To Find the Average of a Set of Numbers
>
> 1. Find the _____
>
> 2. Divide this sum by the _____
>
> **PROCEDURE**

▶ Watch and Work

Watch the video for Example 7 in the software and follow along in the space provided.

Example 7 Calculating an Average

Find the average of the following set of numbers: 15, 8, 90, 35, 27.

Solution

✏️ Now You Try It!

Use the space provided to work out the solution to the next example.

Example A Calculating an Average

Find the average of the following set of numbers: 18, 29, 6, 33, 14, 26.

1.R.4 Exercises

Concept Check

True/False. Determine whether each statement is true or false. If a statement is false, explain how it can be changed so the statement will be true. (**Note:** There may be more than one acceptable change.)

1. Averages are found by performing addition and then division.

2. The sum of 312 and 4 is 1248.

3. The word "quotient" indicates multiplication.

4. After reading a problem carefully, the next step might be to make a diagram or draw a figure.

Applications

Solve.

5. *Nutrition Facts:* Steven is calculating how many calories are in his lunch. He has a hamburger that has 354 calories, a medium fry that has 365 calories, and a chocolate milk shake that has 384 calories. How many total calories is his meal?

6. *Purchases:* For a class in statistics, Anthony bought a new graphing calculator for $95, special graphing paper for $8, a USB flash drive for $10, a textbook for $105, and a workbook for $37. How much did he spend for this class?

7. *Area:* A square that is 10 inches on a side is placed inside a rectangle that has a width of 20 inches and a length of 24 inches. What is the area of the region inside the rectangle that surrounds the square? (Find the area of the shaded region in the figure.)

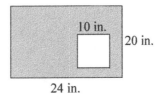

10 in.

20 in.

24 in.

8. *Purchases:* The Lee family spent the following amounts for groceries: $338 in June; $307 in July; $318 in August. What was the average amount they spent for groceries in these three months?

Writing & Thinking

9. Make up three word problems that include key words to indicate operations such as addition, subtraction, multiplication and division. Underline the key words.

10. Give an example where you might use average (other than in a class).

1.R.5 Translating English Phrases and Algebraic Expressions

Key Words To Look For When Translating Phrases

Addition	Subtraction	Multiplication	Division	Exponent (Powers)

1. Division and subtraction are done with the values in the _____ that they are given in the problem.

2. An **ambiguous phrase** is one whose meaning is _____

   _____

▶ Watch and Work

Watch the video for Example 3 in the software and follow along in the space provided.

Example 3 Translating Algebraic Expressions into English Phrases

Change each algebraic expression into an equivalent English phrase. In each case translate the variable as "a number."

a. $5x$

b. $2n + 8$

c. $3(a - 2)$

Solution

✏ Now You Try It!

Use the space provided to work out the solution to the next example.

Example A Identifying Properties of Addition and Multiplication

Change each algebraic expression into an equivalent English phrase.

a. $10x$

b. $4a + 7$

c. $7(n - 5)$

1.R.5 Exercises

Concept Check

True/False. Determine whether each statement is true or false. If a statement is false, explain how it can be changed so the statement will be true. (**Note:** There may be more than one acceptable change.)

1. The order in which the values are given is particularly important when working with subtraction and division problems.

2. "More than" and "increased by" are key phrases specifying the operation of subtraction.

3. Division is indicated by the phrase "five less than a number."

4. Key phrases for parentheses can be used to limit ambiguity in English phrases.

Practice

Write the algebraic expressions described by the English phrases. Choose your own variable.

5. six added to a number

6. twenty decreased by the product of four and a number

7. eighteen less than the quotient of a number and two

Translate each pair of English phrases into algebraic expressions. Notice the differences between the algebraic expressions and the corresponding English phrases.

8. a. six less than a number

   b. six less a number

9. a. six less than four times a number

   b. six less four times a number

Write the algebraic expression described by the English phrase using the given variables.

10. the cost of purchasing a fishing rod and reel if the rod costs x dollars and the reel costs \$8 more than twice the cost of the rod

Translate each algebraic expression into an equivalent English phrase. (There may be more than one correct translation.)

11. $-9x$

12. $\dfrac{9}{x+3}$

Writing & Thinking

13. Explain why translating addition and multiplication problems from English into algebra may be easier than changing subtraction or division problems. (Consider the properties previously studied.)

14. Explain the difference between $5(n + 3)$ and $5n + 3$ when converting from algebra to English.

1.R.6 **Solving Linear Equations: *ax + b = c***

Procedure for Solving Linear Equations that
Simplify to the Form *ax + b = c*

1. Combine _____

2. Use the **addition principle of equality** and _____

   _____

3. Use the _____ and multiply both sides of the

   equation by the reciprocal of the coefficient of the variable (**or** _____

   _____ **itself**). The _____ will become +1.

4. Check your answer by _____

                                                                                    PROCEDURE

▶ **Watch and Work**

Watch the video for Example 2 in the software and follow along in the space provided.

Example 2 Solving Linear Equations of the Form *ax + b = c*

Solve the equation: $-26 = 2y - 14 - 4y$

Solution

✎ Now You Try It!

Use the space provided to work out the solution to the next example.

Example A Solving Linear Equations of the Form *ax* + *b* = *c*

Solve the equation.

$$-18 = 2y - 8 - 7y$$

1.R.6 Exercises

Concept Check

True/False. Determine whether each statement is true or false. If a statement is false, explain how it can be changed so the statement will be true. (**Note:** There may be more than one acceptable change.)

1. If an equation of the form $ax + b = c$ uses decimal or fractional coefficients, the addition and multiplication principles of equality cannot be used.

2. The first step in solving $2x + 3 = 9$ is to add 3 to both sides.

3. To solve an equation that has been simplified to $4x = 12$, you need to multiply both sides by $\frac{1}{4}$, or divide both sides by 4.

4. When solving a linear equation with decimal coefficients, one approach is to multiply both sides in such a way to give integer coefficients before solving.

Practice

Solve each equation.

5. $3x + 11 = 2$

6. $-5x + 2.9 = 3.5$

7. $\dfrac{2}{5} - \dfrac{1}{2}x = \dfrac{7}{4}$

8. $\dfrac{y}{3} - \dfrac{2}{3} = 7$

Applications

Solve.

9. *Music:* The tickets for a concert featuring the new hit band, Flying Sailor, sold out in 2.5 hours. If there were 35,000 tickets sold, solve the equation $35,000 - 2.5x = 0$ to find the number of tickets sold per hour.

10. *Movies:* All snacks (candy, popcorn, and soda) cost \$3.50 each at the local movie theater. Admission tickets cost \$7.50 each. After a long week, Carlos treats himself to a night at the movies. His movie night budget is \$25 and he spends all his movie money. Solve the equation \$3.50x + \$7.50 = \$25.00 to determine how many snacks Carlos can buy.

Writing & Thinking

11. Find the error(s) made in solving each equation and give the correct solution.

a. $\dfrac{1}{3}x + 4 = 9$

$$3 \cdot \dfrac{1}{3}x + 4 = 3 \cdot 9$$

$$x + 4 = 27$$

$$x + 4 - 4 = 27 - 4$$

$$x = 23$$

b. $5x + 3 = 11$

$$(5x - 3) + (3 - 3) = 11 - 3$$

$$2x + 0 = 8$$

$$\dfrac{2x}{2} = \dfrac{8}{2}$$

$$x = 4$$

CHAPTER 2.R

Review Concepts

for *Set Theory*

2.R.1 The Real Number Line and Absolute Value

Integers

The set of numbers consisting of the _____

_____

DEFINITION

Variables

A **variable** is a symbol (generally a _____) that is used to _____

_____

DEFINITION

Rational Numbers

A **rational number** is a number that can be written in

_____

_____

OR

A **rational number** is a number that can be written in _____

_____

DEFINITION

---

1. Numbers that cannot be written as fractions with integer numerators and denominators are

    called _____

▶ Watch and Work

Watch the video for Example 4 in the software and follow along in the space provided.

Example 4 Graphing Sets of Numbers

Graph the set of **real numbers** $\left\{-\dfrac{3}{4}, 0, 1, 1.5, 3\right\}$.

Solution

Now You Try It!

Use the space provided to work out the solution to the next example.

Example A   Graphing Sets of Numbers

Graph the set of real numbers

$$\left\{-2.5, -1, 0, \frac{5}{4}, 4\right\}.$$

Symbols of Equality and Inequality

Reading from left to right:

= _____   ≠ _____

< _____   > _____

≤ _____   ≥ _____

Absolute Value

The **absolute value** of a real number is _____ Note that the
absolute value of a _____

PROPERTIES

2.R.1 **Exercises**

Concept Check

True/False. Determine whether each statement is true or false. If a statement is false, explain how it can be changed so the statement will be true. (**Note:** There may be more than one acceptable change.)

1. On a number line, smaller numbers are always to the left of larger numbers.

2. The absolute value of a negative number is a positive number.

3. All whole numbers are also integers.

4. Zero is a positive number.

Graph each set of real numbers on a real number line.

5. $\{-3, -2, 0, 1\}$

6. $\left\{-2, -1, -\dfrac{1}{3}, 2\right\}$

List the numbers in the set $A = \left\{-7, -\sqrt{6}, -2, -\dfrac{5}{3}, -1.4, 0, \dfrac{3}{5}, \sqrt{5}, \sqrt{11}, 4, 5.9, 8\right\}$ that are described in each exercise.

7. Whole numbers

8. Rational numbers

Determine whether each statement is true or false. If a statement is false, rewrite it in a form that is a true statement. (There may be more than one way to correct a statement.)

9. $0 = -0$

10. $|-8| \geq 4$

Applications

Solve. Represent each quantity with a signed integer.

11. *Oceans:* The Alvin is a manned deep-ocean research submersible that has explored the wreck of the Titanic. The operating depth of the Alvin is 4500 meters below sea level.

12. *Oceans:* The Mariana trench is the deepest known location on the Earth's ocean floor. The deepest known part of the Mariana Trench is approximately 11 kilometers below sea level.

Writing & Thinking

13. Explain, in your own words, how an expression such as $-y$ might represent a positive number.

14. Compare and contrast absolute value with opposites.

2.R.2 **Addition with Real Numbers**

Rules for Addition with Real Numbers

1. To add two real numbers with **like signs**,

    a. _____

    b. use the _____

2. To add two real numbers with **unlike signs**,

    a. _____

    b. use the _____

PROCEDURE

▶ Watch and Work

Watch the video for Example 3 in the software and follow along in the space provided.

Example 3 Adding Three or More Real Numbers

Add.

a. $-3+2+(-5)$

b. $6.0+(-4.3)+(-1.5)$

Solution

✏ Now You Try It!

Use the space provided to work out the solution to the next example.

Example A Adding Three or More Real Numbers

Add.

a. $-7+5+(-3)$

b. $-3.2+(-6.1)+5.7$

2.R.2 **Exercises**

Concept Check

True/False. Determine whether each statement is true or false. If a statement is false, explain how it can be changed so the statement will be true. (**Note:** There may be more than one acceptable change.)

1. The sum of a positive number and a negative number is always positive.

2. When adding two numbers with unlike signs, the result uses the sign of the number with the larger absolute value.

3. The sum of two positive numbers can equal zero.

Practice

Add. Reduce any fractions to lowest terms.

4. $8+(-3)$

5. $2+(-8)$

6. $-\dfrac{1}{6}+\dfrac{7}{15}$

7. $3.2+(-1.2)+(-2.5)$

Add. Be sure to find the absolute values first.

8. $13+|-5|$

Applications

Solve.

9. ***Profit:*** For 2017, a business reports a profit of $45,000 during the first quarter, a loss of $8000 during the second quarter, a loss of $2000 during the third quarter, and a profit of $15,000 during the fourth quarter.

   a. Write an addition expression to represent the total profit made by the company in 2017. Do not simplify.

   b. Simplify the expression from Part **a.**

10. ***Oceans:*** A submarine dives to a depth of 250 feet below the surface. It rises 75 feet before diving an additional 100 feet. What is the final depth of the submarine?

Writing & Thinking

11. Describe, in your own words, how the sum of the absolute values of two numbers might be 0. (Is this even possible?)

12. Describe in your own words the conditions under which the sum of two integers will be 0.

2.R.3 Subtraction with Real Numbers

Additive Inverse

The opposite of a _____. The sum of a number and its

additive inverse _____. Symbolically, for any real number a,

_____

DEFINITION

Subtraction

For any real numbers a and b,

_____

In words, to subtract b from a, _____

DEFINITION

1. To find the **change in value** between two numbers, _____

_____. Symbolically, change in value = _____

_____.

▶ Watch and Work

Watch the video for Example 4 in the software and follow along in the space provided.

Example 4 Application: Calculating Change in Value

A jet pilot flew her plane from an altitude of 30,000 ft to an altitude of 12,000 ft. What was the change in altitude?

Solution

✎ **Now You Try It!**

Use the space provided to work out the solution to the next example.

Example A Calculating Change in Value

A drone plane flew from an altitude of 25,000 ft to an altitude of 14,000 ft. What was the change in altitude?

2.R.3 **Exercises**

True/False. Determine whether each statement is true or false. If a statement is false, explain how it can be changed so the statement will be true. (**Note:** There may be more than one acceptable change.)

1. The sum of a number and its additive inverse is the number itself.

2. The additive inverse of negative seven is seven.

3. We can think of addition of numbers as accumulating numbers.

4. The expression "15 − 7" can be thought of as "fifteen plus negative seven."

Practice

Find the additive inverse (opposite) of each real number.

5. 11

6. −3.4

Subtract. Reduce fractions to lowest terms.

7. $-8-(-11)$

8. $\dfrac{7}{15}-\dfrac{2}{15}$

Perform the indicated operation to find the net change in value.

9. $-6+(-4)-5$

10. $-11.3+5.3-7.9$

Applications

Solve.

11. *Temperature:* At 2 p.m. the temperature was 76 °F. At 8 p.m. the temperature was 58 °F. What was the change in temperature?

12. *Real Estate:* A couple sold their house for $135,000. They paid the realtor $8100, and other expenses of the sale came to $800. If they owed the bank $87,000 for the mortgage, what were their net proceeds from the sale?

Writing & Thinking

13. Explain, in your own words, how to find the difference between a positive and a negative number.

14. What is the additive inverse of 0? Why?

2.R.4 Multiplication and Division with Real Numbers

Rules for Multiplication with Real Numbers

If a and b are positive real numbers, then

1. The product of two positive numbers _____

2. The product of two negative numbers _____

3. The product of a positive number and a negative number _____

4. The product of 0 and any number _____

PROCEDURE

Division with Real Numbers

For real numbers a, b, and x (where $b \neq 0$),

$$\frac{a}{b} = x \text{ means } \underline{\hspace{2cm}}$$

For real numbers a and b (where $b \neq 0$),

$$\frac{a}{0} \text{ is } \underline{\hspace{2cm}}$$

DEFINITION

Rules for Division with Real Numbers

If a and b are positive real numbers (where $b \neq 0$),

1. The quotient of two positive numbers

    _____

2. The quotient of two negative numbers

    _____

3. The quotient of a positive number and a negative number

    _____

PROCEDURE

Average

The **average** (or **mean**) of a set of numbers is the value found by _____

_____

DEFINITION

▶ Watch and Work

Watch the video for Example 6 in the software and follow along in the space provided.

Example 6 Application: Calculating an Average

At noon on five consecutive days in Aspen, Colorado the temperatures were $-5°$, $7°$, $6°$, $-7°$, and $14°$ (in degrees Fahrenheit). (Negative numbers represent temperatures below zero). Find the average of these noonday temperatures.

Solution

✏ Now You Try It!

Use the space provided to work out the solution to the next example.

Example A Calculating an Average

At noon on five consecutive days in Mears, Michigan the temperatures were $-3°$, $5°$, $8°$, $-4°$, and $14°$(in degrees Fahrenheit.) (Negative numbers represent temperatures below zero.) Find the average of these noonday temperatures.

2.R.4 Exercises

True/False. Determine whether each statement is true or false. If a statement is false, explain how it can be changed so the statement will be true. (**Note:** There may be more than one acceptable change.)

1. If a negative number is divided by a positive number, the result will be a negative number.

2. The product of zero and a number is zero.

3. If two numbers have the same sign, both the product and the quotient of the two numbers will be negative.

4. The mean of a set of numbers is always positive.

Practice

Multiply. Reduce fractions to lowest terms.

5. $12 \cdot 4$

6. $(-7)(-16)(0)$

Divide. Reduce fractions to lowest terms. Round answers with decimals to the nearest tenth.

7. $\dfrac{-20}{-10}$

8. $\dfrac{-5.6}{7}$

Applications

Solve.

9. *Mean:* Find the mean of the following set of integers: -10, 15, 16, -17, -34, and -42.

10. *Animals:* According to the US Fish and Wildlife Service, migratory birds are imported at a value of about $19 each. Suppose that about 800,000 live birds are imported each year. What is the total value of these imported birds?

Writing & Thinking

11. If you multiply an odd number of negative numbers together, do you think that the product will be positive or negative? Explain your reasoning.

12. Explain the conditions under which the quotient of two numbers is 0.

2.R.5 Order of Operations with Real Numbers

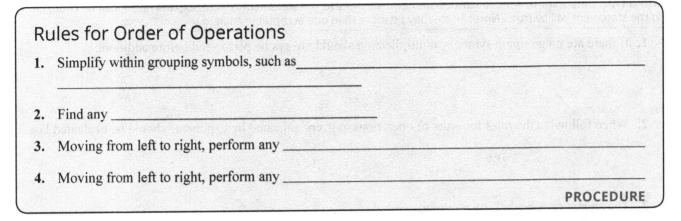

Rules for Order of Operations

1. Simplify within grouping symbols, such as_____

   _____

2. Find any _____

3. Moving from left to right, perform any _____

4. Moving from left to right, perform any _____

PROCEDURE

▶ Watch and Work

Watch the video for Example 3 in the software and follow along in the space provided.

Example 3 Using the Order of Operations with Real Numbers

Simplify: $(2-5)^2 + \left|2-5^2\right| - 2^3$

Solution

✐ Now You Try It!

Use the space provided to work out the solution to the next example.

Example A Using the Order of Operations with Real Numbers

Simplify: $(1-3)^2 + \left|9-4^2\right| - 1^3$

2.R.5 Exercises

True/False. Determine whether each statement is true or false. If a statement is false, explain how it can be changed so the statement will be true. (**Note:** There may be more than one acceptable change.)

1. If there are no grouping symbols, multiplication should always be performed before addition.

2. When following the rules for order of operations, powers indicated by exponents should be evaluated last.

3. The square root symbol is a grouping symbol.

4. A well-known mnemonic device for remembering the rules for order of operations is SADMEP.

Practice

Simplify.

5. a. $24 \div 4 \cdot 6$

   b. $24 \cdot 4 \div 6$

6. $15 \div (-3) \cdot 3 - 10$

7. $3^2 \div (-9) \cdot (4 - 2^2) + 5(-2)$

8. $14 \cdot 3 \div (-2) - 6(4)$

9. $|16 - 20| + (-10)^2 + 5^2$

Applications

Solve.

10. *Discounts:* The Matthews family, a family of 4, is planning a trip to New York City. During their visit, they want to see the Broadway play *Matilda*. The tickets cost $102 each. The Matthews purchase the tickets online and the website charges a service fee of $7.50 per ticket. The website is running a sale where the Matthews can get 10% off of their entire purchase.

    a. Write an expression to describe how much of a discount the Matthews will receive on their purchase.

    b. What is the final purchase price of the tickets?

11. *Banking:* Dennis overdrew his checking account and ended up with a balance of −$42. The bank charged a $35 overdraft fee and an additional $5 fee for every day the account was overdrawn. Dennis left his account overdrawn for 3 days.

    a. Write an expression to show the balance of Dennis's checking account after 3 days.

    b. Simplify the expression in Part **a.** to find the balance of Dennis's checking account after 3 days.

Writing & Thinking

12. Explain, in your own words, why the following expression cannot be evaluated.

$$\left(24 - 2^4\right) + 6\left(3 - 5\right) \div \left(3^2 - 9\right)$$

13. Consider any number between 0 and 1. If you square this number, will the result be larger or smaller than the original number? Is this always the case? Explain.

Review Concepts

for *Rates, Ratios, Proportions, and Percentages*

4.R.1 Introduction to Fractions and Mixed Numbers

1. Numbers such as $\frac{2}{3}$ (read "two-thirds") are said to be in _____.

2. The top number, 2, is called the _____ and the bottom number, 3, is called

   the _____.

Proper Fractions and Improper Fractions

A **proper fraction** is a fraction in which the _____

Examples: _____

An **improper fraction** is a fraction in which the _____

Examples: _____

DEFINITION

Variable

A **variable** is _____

DEFINITION

The Number 0 in Fractions

For any nonzero value of b,

_____

For any value of a,

_____

DEFINITION

---

3. To graph the fraction $\frac{2}{3}$ proceed as follows.

   1. Divide the interval (distance) from _____

   2. Graph (or shade) the _____

3 equal parts

4. A **mixed number** is _____

5. To graph the mixed number $1\frac{3}{4}$ proceed as follows.

1. Mark the intervals from _____

2. Graph (or shade) the
   _____

To Change a Mixed Number to an Improper Fraction

1. Multiply the whole number by _____

2. Add the numerator of the _____

3. Write this sum _____

PROCEDURE

▶ Watch and Work

Watch the video for Example 12 in the software and follow along in the space provided.

Example 12   Changing Mixed Numbers to Improper Fractions

Change $8\frac{9}{10}$ to an improper fraction.

Solution

✏ Now You Try It!

Use the space provided to work out the solution to the next example.

Example A Changing Mixed Numbers to Improper Fractions

Change $10\dfrac{4}{9}$ to an improper fraction.

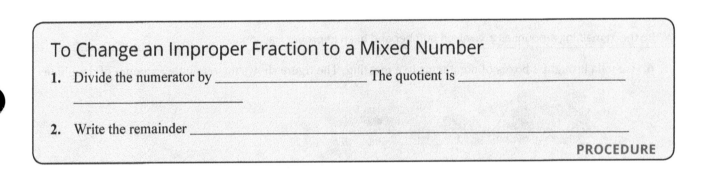

To Change an Improper Fraction to a Mixed Number

1. Divide the numerator by _____ The quotient is _____

    _____

2. Write the remainder _____

PROCEDURE

4.R.1 **Exercises**

Concept Check

True/False. Determine whether each statement is true or false. If a statement is false, explain how it can be changed so the statement will be true. (**Note:** There may be more than one acceptable change.)

1. In $\dfrac{11}{13}$, the denominator is 11.

2. $\dfrac{0}{6} = 0$

3. $\dfrac{17}{0}$ is undefined.

Practice

For the figure, a) write the fraction for the number of days remaining in June (not crossed out) and b) write the fraction for the number of days that have been crossed out for June.

4.

June						
S	M	T	W	T	F	S
~~1~~	~~2~~	~~3~~	~~4~~	~~5~~	~~6~~	~~7~~
~~8~~	~~9~~	~~10~~	~~11~~	~~12~~	~~13~~	~~14~~
~~15~~	~~16~~	~~17~~	~~18~~	~~19~~	~~20~~	~~21~~
~~22~~	~~23~~	24	25	26	27	28
29	30					

5. Graph $\dfrac{3}{5}$ on a number line

Write the remaining amount as **a.** a mixed number and **b.** an improper fraction.

6. Isabella brought 2 boxes of doughnuts to a meeting. The figure shows the remaining amount of doughnuts.

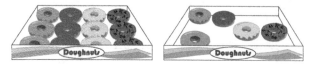

7. Graph $3\dfrac{1}{4}$ on a number line.

8. Change $1\dfrac{3}{5}$ to an improper fraction.

9. Change $\dfrac{4}{3}$ to a mixed number.

Applications

Solve.

10. *Grades:* In a class of 35 students, 6 students received As on a mathematics exam. What fraction of students received an A? What fraction of students did not receive an A?

11. *Nutrition:* A certain brand of plain bagels has 146 calories per bagel. 115 calories come from the carbohydrates in the bagel. What fraction of the calories is from carbohydrates?

Writing & Thinking

12. In your own words, list the parts of a fraction and briefly describe the purpose of each part.

13. Show and explain, using diagrams and words, why $2\dfrac{3}{5} = \dfrac{13}{5}$.

4.R.2 Introduction to Decimal Numbers

To Read or Write a Decimal Number

1. Read (or write) the _____

2. Read (or write) the _____

3. Read (or write) the _____ .

   Then, name the fraction part with the _____ .

PROCEDURE

To Compare Two Decimal Numbers

1. Moving left to right, compare digits _____

2. When one compared digit is larger, then the _____

PROCEDURE

▶ Watch and Work

Watch the video for Example 6 in the software and follow along in the space provided.

Example 6 Comparing Decimal Numbers

Arrange the following three decimal numbers in order from smallest to largest: 6.67, 5.14, 6.28. Then, graph them on a number line.

Solution

✏ Now You Try It!

Use the space provided to work out the solution to the next example.

Example A Comparing Decimal Numbers

Arrange the following three decimal numbers in order from smallest to largest: 2.6, 2.06, 2.46. Then, graph them on a number line.

Rounding Rule for Decimal Numbers

1. Look at the single digit one place value to the right of the digit in the place of desired accuracy.

    a. **If this digit is less than 5,** _____

       and replace all digits to the right with zeros. All digits to the _____.

    b. **If this digit is 5 or greater,** _____

       and replace all digits to the right with zeros. All digits to the _____

       _____. Then, the 9 is replaced by 0 and the _____

       _____

2. Zeros to the right of the place of accuracy that are also to the right of the _____

    _____

                                                                                      PROCEDURE

4.R.2 Exercises

Concept Check

True/False. Determine whether each statement is true or false. If a statement is false, explain how it can be changed so the statement will be true. (**Note:** There may be more than one acceptable change.)

1. Two hundred thousand, four hundred six and twelve hundredths can be written as 200,406.12.

2. 92.586 is greater than 92.6.

3. On a number line, any number to the left of another number is larger than that other number.

4. When a decimal number is rounded, all numbers to the right of the place of accuracy become zeros in the final answer.

Practice

5. Write $2\frac{57}{100}$ in decimal notation.

6. Write 20.7 in words.

7. Write six and twenty-eight thousandths in decimal notation.

8. Arrange 0.2, 0.26, and 0.17 in order from smallest to largest. Then, graph the numbers on a number line.

Fill in the blanks to correctly complete each statement.

9. Round 3.00652 to the nearest ten-thousandth.

   a. The digit in the ten-thousandths position is ___.

   b. The next digit to the right is ___.

   c. Since ___ is less than 5, leave ___ as it is and replace ___ with 0.

   d. So 3.00652 rounds to _____ to the nearest ten-thousandth.

Applications

In each exercise, write the decimal numbers that are not whole numbers in words.

10. *Unicycles:* The tallest unicycle ever ridden was 114.8 feet tall, and was ridden by Sam Abrahams (with a safety wire suspended from an overhead crane) for a distance of 28 feet in Pontiac, Michigan, on January 29, 2004.[1]

11. *Water Weight:* One quart of water weighs approximately 2.0825 pounds.

Writing & Thinking

12. Discuss situations where you think it is particularly appropriate (or necessary) to write numbers in English word form.

13. With **a.** and **b.** as examples, explain in your own words how you can tell quickly when one decimal number is larger (or smaller) than another decimal number.

    a. The decimal number 2.765274 is larger than the decimal number 2.763895.

    b. The decimal number 17.345678 is larger than the decimal number 17.345578.

---

1 Source: http://semcycle.biz/record

4.R.3 Decimals and Percents

The word percent comes from the Latin *per centum*, meaning _____. So **percent means** _____, or **the ratio of a number to** _____.

The symbol % is called the _____. This sign has the same meaning as the fraction $\frac{1}{100}$.

To Change a Decimal Number to a Percent

1. Move the _____

2. Write the _____

PROCEDURE

To Change a Percent to a Decimal Number

1. Move the _____

2. Delete the _____

PROCEDURE

▶ Watch and Work

Watch the video for Example 3 in the software and follow along in the space provided.

Example 3 Changing Percents to Decimal Numbers

Change each percent to a decimal number.

a. 76.%

b. 18.5%

c. 50%

d. 100%

e. 0.25%

✏ Now You Try It!

Use the space provided to work out the solution to the next example.

Example A Changing Percents to Decimal Numbers

Change each percent to a decimal number.

a. 40%

b. 211%

c. 0.6%

d. 29.37%

e. 102%

4.R.3 Exercises

Concept Check

True/False. Determine whether each statement is true or false. If a statement is false, explain how it can be changed so the statement will be true. (**Note:** There may be more than one acceptable change.)

1. If a decimal number is less than 1, then the equivalent percent will be less than 100%.

2. It is not possible to have a percent greater than 100%.

3. A decimal number that is between 0.01 and 0.10 is between 10% and 100%.

4. To change from a percent to a decimal, simply omit the percent sign.

Practice

Change each fraction to a percent.

5. $\dfrac{20}{100}$

6. $\dfrac{125}{100}$

Change each decimal number to a percent.

7. 0.02

8. 2.3

Change each percent to a decimal number.

9. 7%

10. 179%

Applications

Solve.

11. *Interest:* A savings account is offering an interest rate of 0.04 for the first year after opening the account. Change 0.04 to a percent.

12. *Sales Tax:* Suppose that sales tax is figured at 7.25%. Change 7.25% to a decimal.

Writing & Thinking

13. Describe the relationship between percent and the number 100.

14. Describe a situation where more than 100% is possible. Describe a situation where it is impossible to have more than 100%.

4.R.4 Fractions and Percents

> **To Change a Fraction to a Percent**
>
> 1. Change the _____ . (Divide _____ .)
>
> 2. Change the _____ .
>
> **PROCEDURE**

▶ Watch and Work

Watch the video for Example 3 in the software and follow along in the space provided.

Example 3 Changing Mixed Numbers to Percents

Change $2\frac{1}{4}$ to a percent.

Solution

✐ Now You Try It!

Use the space provided to work out the solution to the next example.

Example A Changing Mixed Numbers to Percents

Change $1\frac{1}{2}$ to a percent.

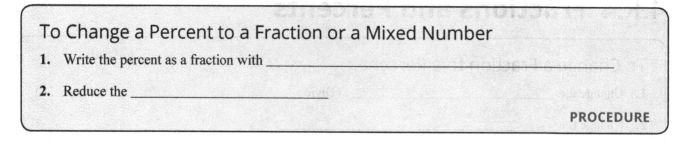

To Change a Percent to a Fraction or a Mixed Number

1. Write the percent as a fraction with _____

2. Reduce the _____

PROCEDURE

4.R.4 Exercises

Concept Check

True/False. Determine whether each statement is true or false. If a statement is false, explain how it can be changed so the statement will be true. (**Note:** There may be more than one acceptable change.)

1. Fractions that have denominators other than 100 cannot be changed to a percent.

2. The fraction $\dfrac{1}{5}$ is equivalent to $\dfrac{1}{5}\%$.

3. When changing from a percent to a mixed number, the fraction does not need to be reduced.

4. $75\% = 0.75 = \dfrac{3}{4}$

Practice

Change each fraction or mixed number to a percent. If necessary, round to the nearest tenth of a percent.

5. $\dfrac{3}{4}$

6. $5\dfrac{3}{10}$

Change each percent to a fraction or mixed number and reduce, if possible.

7. 120%

8. 12.5%

Applications

Solve.

9. *Exam Grades:* Out of a possible total of 240 points on an exam, David received 204 points. What percent of the exam did David get correct?

10. *College Degrees:* To receive a Bachelor of Science (BS) degree at Bluefield State College, the student must complete a total of 128 credit hours, of which 41 of these credits must be general education Core Skills courses. What percent of the total curriculum is dedicated to general education courses? [1]

1   Source: 2010–2011 Bluefield, WV State College Catalogue, p.76

Writing & Thinking

11. Justify why mixed numbers are a larger percentage than proper fractions alone. (Consider the value of 100%.)

12. Describe the process to change a percent to a fraction or mixed number.

4.R.5 Solving Percent Problems Using Proportions

The Percent Proportion $\dfrac{P}{100} = \dfrac{A}{B}$

For the proportion $\dfrac{P}{100} = \dfrac{A}{B}$

$P\%$ = _____ (written _____).

B = _____ (number that _____).

A = _____ (a part of _____).

FORMULA

Three Basic Types of Percent Problems and the Proportion $\dfrac{P}{100} = \dfrac{A}{B}$

Type 1: Find the amount given the _____

What is _____

For example, what is
_____

Type 2: Find the base given the _____

$P\%$ of what _____

For example
_____

Type 3: Find the percent given the _____

What percent of _____

For example, what
_____

FORMULA

▶ Watch and Work

Watch the video for Example 1 in the software and follow along in the space provided.

Example 1 Finding the Amount

What is 65% of 500?

Solution

✏ Now You Try It!

Use the space provided to work out the solution to the next example.

Example A Finding the Amount

What is 15% of 80?

4.R.5 Exercises

Concept Check

True/False. Determine whether each statement is true or false. If a statement is false, explain how it can be changed so the statement will be true. (**Note:** There may be more than one acceptable change.)

1. Percent problems can be solved with a proportion if two of the three parts P, A, and B are known.

2. In the proportion $\dfrac{P}{100} = \dfrac{65}{200}$, the base is 65.

3. In the problem "What is 26% of 720?" the missing number is the base.

4. Because the base represents the whole, it is always larger than the amount.

Practice

Use the proportion $\dfrac{P}{100} = \dfrac{A}{B}$ to find each unknown quantity. Round percents to the nearest tenth of a percent. All other answers should be rounded to the nearest hundredth, if necessary.

5. Find 15% of 50.

6. What is 85% of 60?

7. 25% of 60 is _____.

8. What percent of 48 is 12?

9. _____% of 56 is 140.

Applications

Solve.

10. *Baseball Attendance:* In 2016 the Los Angeles Dodgers led the major leagues in home attendance, drawing an average of 45,720 fans to their home games. This figure represented 81.64% of the capacity of Dodger Stadium. Estimate how many fans the stadium can hold (to the nearest ten) when it is filled to capacity.[1]

11. *Real Estate:* You want to purchase a new home for $122,000. The bank will loan you 80% of the purchase price. How much will the bank loan you? (This amount is called your mortgage and you will pay it off over several years with interest. For example, a 30-year loan will probably cost you a total of more than 3 times the original loan amount.)

Writing & Thinking

12. List the four parts of the proportion equation and give a brief definition of each one.

13. Can a mixed number be used in a proportion? Justify your answer.

1 Source: espn.go.com/mlb/attendance

Review Concepts

for *The Mathematics of Growth*

5.R.1 The Cartesian Coordinate System

1. Descartes based his system on a relationship between _____ in a plane and _____ of real numbers.

2. In the ordered pair (x, y), x is called the _____ and y is called the _____ _____.

3. In an ordered pair of the form (x, y), the _____ is called the **independent variable** and the _____ is called the **dependent variable**.

4. The Cartesian coordinate system relates algebraic equations and ordered pairs to geometry. In this system, two number lines intersect at right angles and separate the plane into four _____. The **origin**, designated by the ordered pair $(0, 0)$, is _____. The horizontal number line is called the _____ or _____. The vertical number line is called the _____ or _____.

One-to-One Correspondence

_____

_____

DEFINITION

▶ Watch and Work

Watch the video for Example 4 in the software and follow along in the space provided.

Example 4 Finding Ordered Pairs

Complete the table so that each ordered pair will satisfy the equation $y = -3x + 1$.

x	y	(x, y)
0		
	4	
$\dfrac{1}{3}$		
3		

Solution

✏ Now You Try It!

Use the space provided to work out the solution to the next example.

Example A Finding Ordered Pairs

Complete the table so that each ordered pair will satisfy the equation $y = -3x + 2$.

x	y	(x, y)
0		
	1	
-2		
	0	

5.R.1 **Exercises**

True/False. Determine whether each statement is true or false. If a statement is false, explain how it can be changed so the statement will be true. (**Note:** There may be more than one acceptable change.)

1. The graph of every ordered pair that has a positive x-coordinate and a negative y-coordinate can be found in Quadrant IV.

2. To find the y-value that corresponds with $x = 2$, substitute 2 for x into the given equation and solve for y.

3. If $(-7, 3)$ is a solution of $y = 3x + 24$, then $(-7, 3)$ satisfies $y = 3x + 24$.

4. If point $A = (0, 4)$, then point A lies on the x-axis.

Practice

List the set of ordered pairs corresponding to the points on the graph.

5.

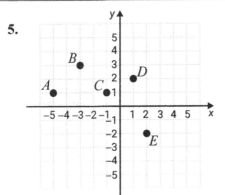

Plot each set of ordered pairs and label the points.

6. $\{A(4, -1), B(3, 2), C(0, 5), D(1, -1), E(1, 4)\}$

Determine the missing coordinate in each of the ordered pairs so that the point will satisfy the equation given.

7. $x - 2y = 2$

   a. $(0, \underline{})$

   b. $(4, \underline{})$

   c. $(\underline{}, 0)$

   d. $(\underline{}, 3)$

Complete the tables so that each ordered pair will satisfy the given equation. Plot the resulting sets of ordered pairs.

8. $y = 2x - 3$

x	y
0	
	−1
−2	
	3

Determine which, if any, of the ordered pairs satisfy the given equation.

9. $2x - 3y = 7$

    a. $(1, 3)$

    b. $\left(\dfrac{1}{2}, -2\right)$

    c. $\left(\dfrac{7}{2}, 0\right)$

    d. $(2, 1)$

The graph of a line is shown. List any three points on the line. (There is more than one correct answer.)

10.

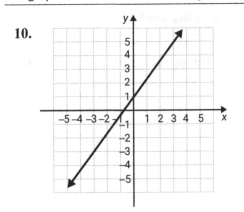

Applications

Solve.

11. ***Exchange Rates:*** At one point in 2017, the exchange rate from US dollars to Euros was $E = 0.85D$ where E is Euros and D is dollars.

   a. Make a table of ordered pairs for the values of D and E if D has the values \$100, \$200, \$300, \$400, and \$500.

   b. Plot the points corresponding to the ordered pairs.

12. ***Temperature:*** Given the equation $F = \frac{9}{5}C + 32$ where C is temperature in degrees Celsius and F is the corresponding temperature in degrees Fahrenheit:

   a. Make a table of ordered pairs for the values of C and F if C has the values $-20°$, $-10°$, $-5°$, $0°$, $5°$, $10°$, and $15°$.

   b. Plot the points corresponding to the ordered pairs.

5.R.2 Graphing Linear Equations in Two Variables

Standard Form of a Linear Equation

Any equation of the form

_____

where A, B, and C are real numbers and A and B are not both equal to 0, is called the standard form of a linear equation.

DEFINITION

To Graph a Linear Equation in Two Variables

1. Locate any two points that _____

2. _____

3. _____

4. To check: Locate a third point that _____

   _____

PROCEDURE

▶ Watch and Work

Watch the video for Example 2 in the software and follow along in the space provided.

_____

Example 2 Graphing a Linear Equation in Two Variables

Graph: $2x + 3y = 6$

Solution

✏ Now You Try It!

Use the space provided to work out the solution to the next example.

Example A Graphing a Linear Equation in Two Variables

Graph: $3x + 2y = 6$

Intercepts

1. To find the y-intercept (where the line crosses the y-axis),

   _____

2. To find the x-intercept (where the line crosses the x-axis),

   _____

PROCEDURE

Horizontal and Vertical Lines

For real numbers a and b, the graph of _____

DEFINITION

5.R.2 Exercises

Concept Check

True/False. Determine whether each statement is true or false. If a statement is false, explain how it can be changed so the statement will be true. (**Note:** There may be more than one acceptable change.)

1. The y-intercept is the point where a line crosses the y-axis.

2. The terms ordered pair and point are used interchangeably.

3. A horizontal line does not have a y-intercept.

4. All x-intercepts correspond to an ordered pair of the form $(0, y)$.

Practice

Graph each linear equation by locating at least two ordered pairs that satisfy the given equation.

5. $x + y = 3$

7. $y = -3$

6. $x = 1$

Graph each linear equation by locating the x-intercept and the y-intercept.

8. $y = 4x - 10$

9. $3x - 7y = -21$

Applications

Solve.

10. ***Chemistry:*** The amount of potassium in a clear bottle of a popular sports drink declines over time when exposed to the UV lights found in most grocery stores. The amount of potassium in a container of this sports drink is given by the equation $y = -30x + 360$, where y represents the mg of potassium remaining after x days on the shelf. Find both the x-intercept and y-intercept, and interpret the meaning of each in the context of this problem.

11. ***Education:*** Mr. Adler has found that the grade each student gets in his Introductory Algebra course directly correlates with the amount of time spent doing homework, and is represented by the equation $y = 7x + 30$, where y represents the numerical score the student receives on an exam (out of 100 points) after spending x hours per week doing homework. Find the y-intercept and interpret its meaning in this context.

Writing & Thinking

12. Explain, in your own words, why it is sufficient to find the x-intercept and y-intercept to graph a line (assuming that they are not the same point).

13. Explain, in your own words, how you can determine if an ordered pair is a solution to an equation.

5.R.3 **Rules for Exponents**

The Product Rule for Exponents

If a is a nonzero real number and m and n are integers, then

$$\underline{\hspace{4cm}}$$

In words, to multiply powers with the same base, _____

PROPERTIES

The Exponent 0

If a is a nonzero real number, then _____

The expression _____ is _____

DEFINITION

Quotient Rule for Exponents

If a is a nonzero real number and m and n are integers, then

$$\underline{\hspace{4cm}}$$

In words, to divide two powers with the same base, _____

PROPERTIES

Rule for Negative Exponents

If a is a nonzero real number and n is an integer, then

$$\underline{\hspace{4cm}}$$

PROPERTIES

▶ Watch and Work

Watch the video for Example 6 in the software and follow along in the space provided.

Example 6 Negative Exponents

Use the rule for negative exponents to simplify each expression so that it contains only positive exponents.

 a. 5^{-1}

 b. x^{-3}

 c. $x^{-9} \cdot x^{7}$

Solution

✎ Now You Try It!

Use the space provided to work out the solution to the next example.

Example A Negative Exponents

Use the rule for negative exponents to simplify each expression so that it contains only positive exponents.

 a. 7^{-1}

 b. x^{-7}

 c. $x^{-11} \cdot x^{6}$

Summary of the Rules for Exponents

For any nonzero real number a and integers m and n:

1. The exponent 1: _____

2. The exponent 0: _____

3. The product rule:_____

4. The quotient rule:
    _____

5. Negative exponents:
    _____

PROPERTIES

5.R.3 Exercises

Concept Check

True/False. Determine whether each statement is true or false. If a statement is false, explain how it can be changed so the statement will be true. (**Note:** There may be more than one acceptable change.)

1. If a constant does not have an exponent written, it is assumed that the exponent is 0.

2. If a is a nonzero real number and n is an integer, then $a^{-n} = -a^{n}$.

3. Since the product rule is stated for integer exponents, the rule is also valid for 0 and negative exponents.

4. When using the quotient rule, you should subtract the smaller exponent from the larger exponent.

Practice

Simplify each expression. The final form of the expressions with variables should contain only positive exponents. Assume that all variables represent nonzero numbers.

5. $y^3 \cdot y^8$

6. $\dfrac{y^7}{y^2}$

7. $x^{-3} \cdot x^0 \cdot x^2$

8. $\dfrac{10^4 \cdot 10^{-3}}{10^{-2}}$

Applications

Solve.

9. *Computers:* Rylee wants to move all her files to a new hard drive that has 2^{12} GB of storage on it. She wants to designate the same amount of storage for each of 2^4 projects. How much storage should be assigned to each project? Write your answer as a power of two.

10. *Bacteria:* Trey is studying patterns in bacteria. For a positive test result in his experiment, bacteria must grow in population at a minimum rate of 3^2 in 24 hours. If the initial population of the bacteria is 3^5 and his final measurement after 24 hours is 3^8, should he mark the test as positive or negative?

5.R.4 Greatest Common Factor (GCF) of a Set of Terms

1. The **greatest common factor (GCF)** of two or more integers is the _____

   _____

Procedure for Finding the GCF of a Set of Terms

1. Find the prime factorization of _____

2. List all the factors that are _____

3. Raise each common factor to the _____

4. Multiply these powers to _____

Note: If there is no common prime factor or variable, then _____

PROCEDURE

▶ Watch and Work

Watch the video for Example 1 in the software and follow along in the space provided.

Example 1 Finding the GCF

Find the GCF for each set of algebraic terms.

a. $\{30, 45, 75\}$

b. $\{20x^4y, 15x^3y, 10x^5y^2\}$

Solution

✏ Now You Try It!

Use the space provided to work out the solution to the next example.

Example A Factoring Polynomials by Grouping

Find the GCF for each set of algebraic terms.

a. $\{5, 10, 20\}$

b. $\{150xy, 250x^3y^2, 100x^2y^2\}$

5.R.4 Exercises

Concept Check

True/False. Determine whether each statement is true or false. If a statement is false, explain how it can be changed so the statement will be true. (**Note:** There may be more than one acceptable change.)

1. When finding the GCF of a polynomial, you need to consider only the coefficients.

2. One way to find the GCF of a set of numbers is to use the prime factorization of each number.

Practice

Find the GCF for each set of terms.

3. $\{25, 30, 75\}$

4. $\{8a^3, 16a^4, 20a^2\}$

5. $\{26ab^2, 39a^2b, 52a^2b^2\}$

6. $\{45x^2y^2z^2, 75xy^2z^3\}$

Applications

Solve.

7. *Halloween:* Bonnie volunteers to bring bags of candy to her child's class for the Halloween party this year. She buys one bag of candy A containing 150 pieces of candy, one bag of candy B containing 180 pieces of candy, and one bag of candy C containing 330 pieces of candy. She needs to use all the candy to create identical treat bags. How many treat bags can Bonnie make so that each one has the same number and variety of candy? How many of each type of candy will be in each bag?

Writing & Thinking

8. Explain why the GCF of $-3x^2 + 3$ is 3 and not -3.

5.R.5 Factoring Trinomials: $x^2 + bx + c$

▶ Watch and Work

Watch the video for Example 1 in the software and follow along in the space provided.

Example 1 Factoring Trinomials with Leading Coefficients of 1

Factor: $x^2 + 8x + 12$

Solution

✏ Now You Try It!

Use the space provided to work out the solution to the next example.

Example A Factoring Trinomials with Leading Coefficients of 1

Factor: $x^2 + 10x + 21$

To Factor Trinomials of the Form $x^2 + bx + c$

To factor $x^2 + bx + c$, if possible, find _____

1. If c is positive, then _____
    a. Both will be _____

        Example: _____
    b. Both will be_____

        Example: _____

2. If c is negative, _____

    Examples: $x^2 + 6x - 7 = (x+7)(x-1)$ and _____

PROCEDURE

5.R.5 Exercises

Concept Check

True/False. Determine whether each statement is true or false. If a statement is false, explain how it can be changed so the statement will be true. (**Note:** There may be more than one acceptable change.)

1. In a trinomial such as $x^2 - 5x + 4$, one would need to find two factors of 4 whose sum is negative 5.

2. In factoring a trinomial with leading coefficient 1, if the constant term is negative, then both factors must be negative.

3. The first step in factoring a trinomial is to look for a common monomial factor.

4. For a trinomial with leading coefficient 1, if no pair exists whose product is the constant and whose sum is the middle term's coefficient, then the trinomial is not factorable.

Practice
Completely factor each trinomial. If a trinomial cannot be factored, write "not factorable."

5. $x^2 - 6x - 27$

6. $a^2 + a + 2$

7. $y^2 - 14y + 24$

8. $2a^4 + 24a^3 + 54a^2$

Applications
Solve.

9. *Triangles:* The area of a triangle is $\dfrac{1}{2}$ the product of its base and its height. If the area of the triangle shown is given by the function $A(x) = \dfrac{1}{2}x^2 + 24x$, find representations for the lengths of its base and its height (where the base is longer than the height).

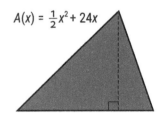

10. *Rectangles:* The area of the rectangle shown is given by the polynomial function $A(x) = 4x^2 + 20x$. If the width of the rectangle is 4x, what is the length?

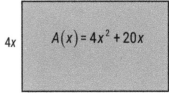

Writing & Thinking

11. Discuss, in your own words, how the sign of the constant term determines what signs will be used in the factors when factoring trinomials.

5.R.6 Factoring Trinomials: $ax^2 + bx + c$

Guidelines for the Trial-and-Error Method

1. If the sign of the constant term is positive (+), the signs in _____
_____

2. If the sign of the constant term is negative (−), the signs in _____
_____

▶ Watch and Work

Watch the video for Example 2 in the software and follow along in the space provided.

Example 2 Factoring Trinomials

Completely factor each polynomial. Be sure to look first for the greatest common monomial factor.

a. $6x^3 - 8x^2 + 2x$

b. $-2x^2 - x + 6$

c. $10x^3 + 5x^2 + 5x$

Solution

✏ Now You Try It!

Use the space provided to work out the solution to the next example.

Example A Using the *ac*-Method

Compeletely factor each polynomial.

a. $8x^3 - 12x^2 + 4x$

b. $21x^3 + 49x^2 - 7x$

5.R.6 Exercises

Concept Check

True/False. Determine whether each statement is true or false. If a statement is false, explain how it can be changed so the statement will be true. (**Note:** There may be more than one acceptable change.)

1. A trinomial is factorable if the middle term is the difference of the inner and outer products of two binomials.

2. The trial-and-error method of factoring a trinomial follows the same steps as the FOIL method of multiplication.

3. Factoring can be checked by multiplying the factors and verifying that the product matches the original polynomial.

Practice

Completely factor each polynomial. If a polynomial cannot be factored, write "not factorable."

4. $6x^2 + 11x + 5$

5. $-x^2 + 3x - 2$

6. $x^2 + 8x + 64$

7. $9x^2 - 3x - 20$

8. $12x^2 - 38x + 20$

9. $5a^2 - 7a + 2$

Writing & Thinking

10. It is true that $2x^2 + 10x + 12 = (2x+6)(x+2) = (2x+4)(x+3)$. Explain how the trinomial can be factored in two ways. Is there some kind of error?

11. It is true that $5x^2 - 5x - 30 = (5x-15)(x+2)$. Explain why this is not the completely factored form of the trinomial.

5.R.7 Special Factoring Techniques

Difference of Two Squares

Consider the polynomial $x^2 - 25$. By recognizing this expression as the **difference of two squares**, we can go directly to the factors:

$$x^2 - 25 = \underline{\hspace{4cm}}$$

DEFINITION

Sum of Two Squares

The sum of two squares is an expression of the form _____

DEFINITION

In a perfect square trinomial, both the _____ and _____ terms of the trinomial must be perfect squares. If the first term is of the form x^2 and the last term is of the form a^2, then the middle term must be of the form _____ or _____.

▶ Watch and Work

Watch the video for Example 3 in the software and follow along in the space provided.

Example 3 Factoring Perfect Square Trinomials

Factor completely.

a. $z^2 - 12z + 36$

b. $4y^2 + 12y + 9$

c. $2x^3 - 8x^2 y + 8xy^2$

d. $\left(x^2 + 6x + 9\right) - y^2$

Solution

✏ Now You Try It!

Use the space provided to work out the solution to the next example.

Example A Factoring Perfect Square Trinomials

Factor completely.

a. $z^2 + 40z + 400$

b. $y^2 - 14y + 49$

c. $3x^2z - 18xyz + 27y^2z$

d. $(y^2 + 8y + 16) - z^2$

5.R.7 **Exercises**

Concept Check

True/False. Determine whether each statement is true or false. If a statement is false, explain how it can be changed so the statement will be true. (**Note:** There may be more than one acceptable change.)

1. The expression $x^2 + 20x + 100$ is a perfect square trinomial.

2. When factoring polynomials, always look for a common monomial factor first.

3. The sum of two squares, $(x^2 + a^2)$, is factorable.

Practice

Completely factor each of the given polynomials. If a polynomial cannot be factored, write "not factorable."

4. $25 - z^2$

5. $y^2 - 16y + 64$

6. $x^2 + 64y^2$

Solve.

7. **a.** Represent the area of the shaded region of the square shown below as the difference of two squares.

   b. Use the factors of the expression in Part **a.** to draw (and label the sides of) a rectangle that has the same area as the shaded region.

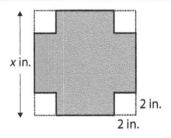

Writing & Thinking

8. a. Show that the sum of the areas of the rectangles and squares in the figure is a perfect square trinomial.

b. Rearrange the rectangles and squares in the form of a square and represent its area as the square of a binomial.

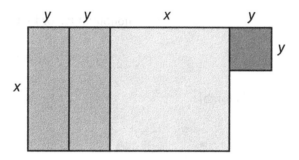

5.R.8 Quadratic Equations: The Quadratic Formula

The Quadratic Formula

For the general quadratic equation _____

_____

FORMULA

▶ Watch and Work

Watch the video for Example 2 in the software and follow along in the space provided.

Example 2 The Quadratic Formula

Solve by using the quadratic formula.

$7x^2 - 2x + 1 = 0$

Solution

✏ Now You Try It!

Use the space provided to work out the solution to the next example.

Example A The Quadratic Formula

Solve by using the quadratic formula.

$5x^2 - 3x + 2 = 0$

---

1. The expression $b^2 - 4ac$, the part of the quadratic formula that lies under the radical sign, is called

   the _____.

Discriminant	Nature of Solutions
$b^2 - 4ac > 0$	_____
$b^2 - 4ac = 0$	_____
$b^2 - 4ac < 0$	_____

Table 1

5.R.8 Exercises

Concept Check

True/False. Determine whether each statement is true or false. If a statement is false, explain how it can be changed so the statement will be true. (**Note:** There may be more than one acceptable change.)

1. The quadratic formula will always work when solving quadratic equations.

2. If the discriminant is a perfect square, the quadratic equation is factorable.

3. When using the quadratic formula, if the discriminant is greater than zero, there are infinite solutions.

4. If the discriminant is less than zero, there is no real solution.

Practice

Find the discriminant and determine the nature of the solutions of each quadratic equation.

5. $x^2 + 6x - 8 = 0$

6. $x^2 - 8x + 16 = 0$

Solve each of the quadratic equations using the quadratic formula.

7. $x^2 + 4x - 4 = 0$

8. $x^2 - 2x + 7 = 0$

9. $3x^2 - 7x + 4 = 0$

Applications

Solve.

10. ***Throwing Objects:*** 🖩 An orange is thrown down from the top of a building that is 300 feet tall with an initial velocity of 6 feet per second. The distance of the object from the ground can be calculated using the equation $d = 300 - 6t - 16t^2$, where t is the time in seconds after the orange is thrown.

   a. On a balcony, a cup is sitting on a table located 100 feet from the ground. If the orange is thrown with the right aim to fall into the cup, how long will the orange fall? Round to the nearest hundredth. (**Hint:** The distance is 100 feet.)

   b. If the orange misses the cup and falls to the ground, how long will it take for the orange to splatter on the sidewalk? (**Hint:** What is the height of the orange when it hits the ground?)

   c. Approximately how much longer would it take for the orange to fall to the sidewalk than it would for the orange to fall into the cup?

Writing & Thinking

11. Find an equation of the form $Ax^4 + Bx^2 + C = 0$ that has the four roots ± 2 and ± 3. Explain how you arrived at this equation.

12. The surface area of a right circular cylinder can be found using the following formula: $S = 2\pi r^2 + 2\pi rh$, where r is the radius of the cylinder and h is the height. Estimate the radius of a circular cylinder of height 30 cm and surface area 300 cm². Explain how you used your knowledge of quadratic equations.

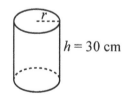

$h = 30$ cm

Review Concepts

for *Geometry*

6.R.1 **Proportions**

Proportions

A **proportion** is a statement that _____

In symbols,

_____

A proportion is true if _____

DEFINITION

To Solve a Proportion

1. Find the cross products (or cross multiply) and then _____

2. Divide both sides of the equation by _____

3. _____

PROCEDURE

▶ **Watch and Work**

Watch the video for Example 3 in the software and follow along in the space provided.

Example 3 Solving Proportions

Find the value of x if $\dfrac{4}{8} = \dfrac{5}{x}$.

Solution

✎ Now You Try It!

Use the space provided to work out the solution to the next example.

Example A Solving Proportions

Find the value of x if $\dfrac{12}{x} = \dfrac{9}{15}$.

To Solve an Application Using a Proportion

1. Identify the unknown quantity and _____

2. Set up a proportion in which the _____. (Make sure that
   the _____)

3. Solve the _____

PROCEDURE

6.R.1 Exercises

Concept Check

True/False. Determine whether each statement is true or false. If a statement is false, explain how it can be changed so the statement will be true. (**Note:** There may be more than one acceptable change.)

1. A proportion is a statement that two ratios are being multiplied.

2. Cross canceling is used to determine if a proportion is true.

3. In order to solve the proportion $\dfrac{16}{36.8} = \dfrac{x}{27.6}$ we construct the equation $36.8x = 441.6$.

4. When using proportions to solve a word problem, there is only one correct way to set up the proportion.

5. The proportions $\dfrac{36 \text{ tickets}}{\$540} = \dfrac{x \text{ tickets}}{\$75}$ and $\dfrac{x \text{ tickets}}{36 \text{ tickets}} = \dfrac{\$75}{\$540}$ will yield the same answer.

Practice

Determine whether each proportion is true or false.

6. $\dfrac{3}{6} = \dfrac{4}{8}$

7. $\dfrac{1}{3} = \dfrac{33}{100}$

Solve each proportion.

8. $\dfrac{5}{4} = \dfrac{x}{8}$

9. $\dfrac{3.5}{2.6} = \dfrac{10.5}{B}$

Applications

Solve.

10. *Concrete:* The quality of concrete is based on the ratio of bags of cement to cubic yards of gravel. One batch of concrete consists of 27 bags of cement mixed into 9 cubic yards of gravel, while a second has 15 bags of cement mixed with 5 cubic yards of gravel. Determine whether the ratio of cement to gravel is the same for both batches.

11. *Grading:* An English teacher must read and grade 27 essays. If the teacher takes 20 minutes to read and grade 3 essays, how much time will he need to grade all 27 essays?

Writing & Thinking

12. In your own words, clarify how you can know that a proportion is set up correctly or not.

6.R.2 Square Roots and the Pythagorean Theorem

> **Terminology of Radicals**
>
> The symbol $\sqrt{}$ is called _____
>
> The number under the radical sign is called _____
>
> The complete expression, such as $\sqrt{49}$, is called _____
>
> **DEFINITION**

▶ Watch and Work

Watch the video for Example 2 in the software and follow along in the space provided.

Example 2 Evaluating Square Roots

Use your memory of the values in Table 2 to evaluate each expression.

a. $\sqrt{256}$

b. $\sqrt{81}$

Solution

✏ Now You Try It!

Use the space provided to work out the solution to the next example.

Example A Evaluating Square Roots

Evaluate each expression.

    a. $\sqrt{36}$

    b. $\sqrt{169}$

Terms Related to Right Triangles

Right triangle: _____

Hypotenuse: _____

Leg: _____

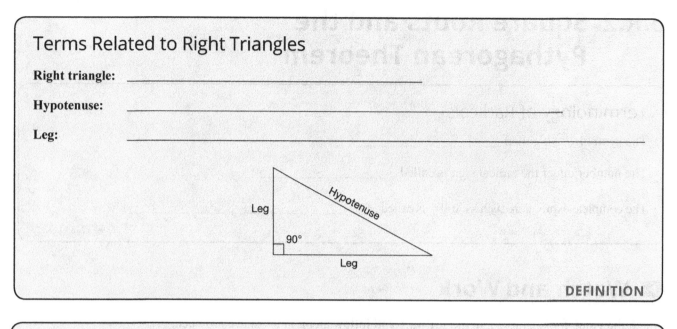

DEFINITION

The Pythagorean Theorem

In a right triangle, the _____

_____

$$\underline{} = \underline{} + \underline{}$$

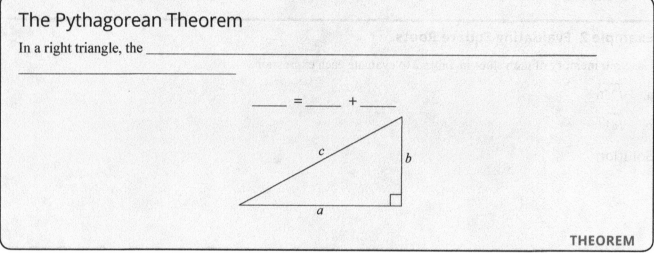

THEOREM

6.R.2 Exercises

Concept Check

True/False. Determine whether each statement is true or false. If a statement is false, explain how it can be changed so the statement will be true. (**Note:** There may be more than one acceptable change.)

1. 49 is a perfect square.

2. In the expression $\sqrt{81}$, the number 9 is the radicand.

3. The Pythagorean Theorem can be used to find the length of the longest side of a right triangle if the lengths of the two legs are known.

4. The Pythagorean Theorem works for any type of triangle.

Practice

Evaluate each expression.

5. $\sqrt{36}$

6. $\sqrt{225}$

Use the Pythagorean Theorem to determine whether or not each triangle is a right triangle.

7.

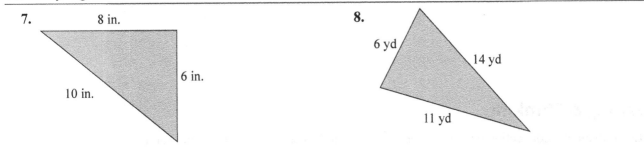

8 in.

6 in.

10 in.

8.

6 yd

14 yd

11 yd

Find the hypotenuse for each right triangle accurate to the nearest hundredth.

9.

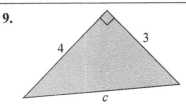

4

3

c

Applications

📖 Solve.

10. *Safety:* The base of a fire engine ladder is 30 feet from a building and reaches to a third floor window 50 feet above ground level. Find the length of the ladder to the nearest hundredth of a foot.

11. *Baseball:* The shape of home plate in the game of baseball can be created by cutting off two triangular pieces at the corners of a square, as shown in the figure. If each of the triangular pieces has a hypotenuse of 12 inches and legs of equal length, what is the length of one side of the original square, to the nearest tenth of an inch?

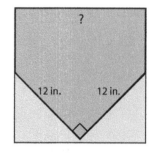

Writing & Thinking

12. Explain the connection between a perfect square and its square root. Give an example.

6.R.3 Simplifying Algebraic Expressions

> ## Combining Like Terms
>
> To combine like terms, _____
>
> **DEFINITION**

▶ Watch and Work

Watch the video for Example 1 in the software and follow along in the space provided.

Example 1 Combining Like Terms

Simplify each expression by combining like terms.

a. $8x + 10x$

b. $6.5y - 2.3y + 3$

c. $2x^2 + 3x + x^2 - x$

d. $4(n - 7) + 5(n + 1)$

e. $3a - 52a^2 + 9$

f. $\dfrac{x + 3x}{2} + 5x$

Solution

Now You Try It!

Use the space provided to work out the solution to the next example.

Example A Identifying the Properties of Addition and Multiplication

Simplify each expression by combining like terms.

   a. $3x + 9x$

   b. $4.2z + 5 - 3.1z$

   c. $4x^2 + 7y + 3x^2 - 2y$

   d. $6(a - 2) + 4(a + 5)$

   e. $-2x^2 + 5y^2 + 9 - y$

   f. $\dfrac{x + 5x}{3} + 6x$

6.R.3 **Exercises**

Concept Check

True/False. Determine whether each statement is true or false. If a statement is false, explain how it can be changed so the statement will be true. (**Note:** There may be more than one acceptable change.)

1. A variable that does not appear to have an exponent has an exponent of 1.

2. Like terms have the same coefficients.

Practice

Simplify each expression by combining like terms.

3. $8x + 7x$

4. $3x - 5x + 12x$

5. $13x + 12x^2 + 15x - 35 - 41 - 2x^2$

6. $5y + 4 - 2y$

7. $\dfrac{3a + 5a}{-2} + 12a$

Applications

Solve.

8. ***Inventory:*** A moving company starts the week with 72 bundles of small boxes and 50 bundles of medium boxes. During the week, they use 25 bundles of small boxes and 32 bundles of medium boxes. At the end of the week, they buy 125 bundles of medium boxes. The total number of boxes at the end of the week can be modeled by the expression $72s + 50m - 25s - 32m + 125m$, where s represents the number of boxes in a bundle of small boxes and m represents the number of boxes in a bundle of medium boxes. Simplify the expression for the total number of boxes at the end of the week by combining like terms.

6.R.4 Evaluating Algebraic Expressions

To Evaluate an Algebraic Expression

1. _____

2. _____

3. _____

(**Note:** Terms separated by + and − signs may be evaluated _____
_____.)

PROCEDURE

▶ Watch and Work

Watch the video for Example 5 in the software and follow along in the space provided.

Example 5 Simplifying and Evaluating Algebraic Expressions

Simplify and evaluate $3ab - 4ab + 6a - a$ for $a = 2$, $b = -1$.

Solution

✏ Now You Try It!

Use the space provided to work out the solution to the next example.

Example A Identifying Properties of Addition and Multiplication

Simplify and evaluate

$5ab - 8ab + 2a - 3a$ for $a = -3$, $b = 1$

6.R.4 Exercises

Concept Check

True/False. Determine whether each statement is true or false. If a statement is false, explain how it can be changed so the statement will be true. (**Note:** There may be more than one acceptable change.)

1. A variable that does not appear to have an exponent has an exponent of 1.

2. Like terms have the same coefficients.

Practice

Simplify each expression and then evaluate the expression for $y = 3$ and $a = -2$.

3. $5y + 4 - 2y$

4. $\dfrac{3a + 5a}{-2} + 12a$

Applications

Solve.

5. *Profit:* An apartment management company owns a property with 100 units. The company has determined that the profit made per month from the property can be calculated using the equation $P = -10x^2 + 1500x - 6000$, where x is the number of units rented per month. How much profit does the company make when 80 units are rented?

6. *Physics:* A ball is thrown upward from an initial height of 96 feet with an initial velocity of 16 feet per second. After t seconds, the height of the ball can be described by the expression $-16t^2 + 16t + 96$. What is the height of the ball after 3 seconds?

Writing & Thinking

7. Discuss like and unlike terms and give an example of each.

8. Explain the difference between -13^2 and $(-13)^2$.

6.R.5 Working with Formulas

1. **Formulas** are general rules or _____ stated _____.

2. We say that the formula $d = rt$ is _____ d _____ r and t. Similarly, the formula $A = \frac{1}{2}bh$ is solved for _____ in terms of _____, and the formula $P = R - C$ (profit is equal to revenue minus cost) is solved for _____ in terms of _____ _____.

▶ Watch and Work

Watch the video for Example 6 in the software and follow along in the space provided.

Example 6 Solving for Different Variables

Given $V = \dfrac{k}{P}$, solve for P in terms of V and k.

Solution

✎ Now You Try It!

Use the space provided to work out the solution to the next example.

Example A Solving for Different Variables

Given $P = \dfrac{I}{rt}$ solve for t in terms of $I, r,$ and P.

6.R.5 Exercises

Concept Check

True/False. Determine whether each statement is true or false. If a statement is false, explain how it can be changed so the statement will be true. (**Note:** There may be more than one acceptable change.)

1. When using formulas, typically it does not matter if capital or lower case letters are used: $A = a, C = c,$ etc.

2. If the perimeter and length are known, $P = 2l + 2w$ can be used to find the width of a rectangle.

3. Rate of interest is stated as an annual rate in percent form.

Applications

In the following application problems, read the descriptions carefully and then substitute the values given in the problem for the corresponding variables in the formulas. Evaluate the resulting expression for the unknown variable.

Velocity

If an object is shot upward with an initial velocity v_0 in feet per second, the velocity v in feet per second is given by the formula $v = v_0 - 32t$, where t is time in seconds. (v_0 is read "v sub zero." The $_0$ is called a subscript.)

4. An object projected upward with an initial velocity of 106 feet per second has a velocity of 42 feet per second. How many seconds have passed?

Investments

The total amount of money in an account with P dollars invested in it is given by the formula $A = P + Prt$, where r is the rate expressed as a decimal and t is time (one year or part of a year).

5. If $1000 is invested at 6% interest, find the total amount in the account after 6 months.

Cost

The total cost C of producing x items can be found by the formula $C = ax + k$, where a is the cost per item and k is the fixed costs (rent, utilities, and so on).

6. Find the total cost of producing 30 items if each costs $15 and the fixed costs are $580.

Solve each formula for the indicated variable.

7. $P = a + b + c$; solve for b.

8. $P = 3s$; solve for s.

9. $I = Prt$; solve for t.

10. $A = P(1 + rt)$; solve for r.

CHAPTER 7.R

Review Concepts

for *Probability*

7.R.1 Multiplication and Division with Fractions and Mixed Numbers

---

To Multiply Fractions

1. _____

2. _____

$$\frac{a}{b} \cdot \frac{c}{d} = \underline{\hspace{1cm}} \quad (b, d \neq \underline{\hspace{0.5cm}})$$

For example, _____

PROCEDURE

Commutative Property of Multiplication

The order of the fractions being multiplied can be _____

$$\frac{a}{b} \cdot \frac{c}{d} = \underline{\hspace{1cm}} \quad (b, d \neq \underline{\hspace{0.5cm}})$$

For example, _____

PROPERTIES

Associative Property of Multiplication

The grouping of the fractions being multiplied can be _____

$$\left(\frac{a}{b} \cdot \frac{c}{d}\right) \cdot \frac{e}{f} = \underline{\hspace{1.5cm}} \quad (b, d, f \neq \underline{\hspace{0.5cm}})$$

For example, _____

PROPERTIES

---

1. **A fraction is reduced to lowest terms if the numerator and denominator** _____

_____

To Reduce a Fraction to Lowest Terms

1. Factor the _____

2. Use the fact that _____

Note: Reduced fractions may be improper fractions.

PROCEDURE

▶ Watch and Work

Watch the video for Example 8 in the software and follow along in the space provided.

Example 8 Multiplying and Reducing Using Prime Factors

Multiply and reduce to lowest terms: $\dfrac{17}{50} \cdot \dfrac{25}{34} \cdot 8$

Solution

✏ Now You Try It!

Use the space provided to work out the solution to the next example.

Example A Multiplying and Reducing Using Prime Factors

Multiply and reduce to lowest terms:

$16 \cdot \dfrac{12}{100} \cdot \dfrac{5}{36}$

Reciprocals

The reciprocal of $\dfrac{a}{b}$ is _____ . The product of a nonzero number and its reciprocal is

_____

$$\frac{a}{b} \cdot \frac{b}{a} = 1$$

Note: $0 = \dfrac{0}{1}$, but $\dfrac{1}{0}$ _____

DEFINITION

To Divide Fractions

To divide by any nonzero number, _____

$$\frac{a}{b} \div \frac{c}{d} = $$

_____

For example, _____

_____

PROCEDURE

To Multiply Mixed Numbers

1. Change each mixed number to _____

2. Factor the numerator and denominator of _____

3. Change the answer to _____

PROCEDURE

2. The area of a triangle is $\frac{1}{2}$ _____

To Divide with Mixed Numbers

1. Change each mixed number to _____

2. _____

3. _____

PROCEDURE

▶ **Watch and Work**

Watch the video for Example 25 in the software and follow along in the space provided.

Example 25   Dividing and Reducing Mixed Numbers

Divide and reduce to lowest terms: $3\dfrac{1}{4} \div 7\dfrac{4}{5}$

Solution

✎ **Now You Try It!**

Use the space provided to work out the solution to the next example.

Example A   Dividing and Reducing Mixed Numbers

Divide and reduce to lowest terms: $2\dfrac{2}{5} \div 3\dfrac{3}{4}$

7.R.1 **Exercises**

Concept Check

True/False. Determine whether each statement is true or false. If a statement is false, explain how it can be changed so the statement will be true. (**Note:** There may be more than one acceptable change.)

1. To find $\frac{1}{2}$ of $\frac{2}{9}$ requires multiplication.

2. The statement $\frac{1}{3} \cdot \frac{2}{5} = \frac{2}{5} \cdot \frac{1}{3}$ is an example of the associative property of multiplication.

3. The number 1 is always a factor of the numerator and the denominator.

4. The reciprocal of 1 is undefined.

5. The product of a nonzero number and its reciprocal is undefined.

6. Multiplication or division with mixed numbers can be accomplished by changing the mixed numbers to improper fractions.

Practice

Multiply and reduce to lowest terms. (**Hint:** Factor before multiplying.)

7. $\frac{5}{16} \cdot \frac{16}{15}$

8. $\frac{9}{10} \cdot \frac{35}{40} \cdot \frac{25}{15}$

Divide and reduce to lowest terms.

9. $0 \div \dfrac{5}{6}$

10. $\dfrac{5}{6} \div 0$

11. $\dfrac{14}{15} \div \dfrac{21}{25}$

Multiply and reduce to lowest terms. Write your answer in mixed number form.

12. $12\dfrac{1}{2} \cdot 3\dfrac{1}{3}$

Divide and reduce to lowest terms. Write your answer in mixed number form.

13. $3\dfrac{1}{2} \div \dfrac{7}{8}$

14. $7\dfrac{1}{5} \div 3$

Applications

Solve.

15. **_Recipes:_** A recipe calls for $\frac{3}{4}$ cups of flour. How much flour should be used if only half of the recipe is to be made?

16. **_Demographics:_** A study showed that $\frac{3}{5}$ of the students in an elementary school were left-handed. If the school had an enrollment of 600 students, how many were left-handed?

17. *Geology:* The floor of the Atlantic Ocean is spreading apart at an average rate of $\frac{3}{50}$ of a meter per year. How long will it take for the ocean floor to spread 12 meters?

18. *Airplane Capacity:* An airplane is carrying 180 passengers. This is $\frac{9}{10}$ of the capacity of the airplane.

    a. Is the capacity of the airplane more or less than 180?

    b. If you were to multiply 180 times $\frac{9}{10}$, would the product be more or less than 180?

    c. What is the capacity of the airplane?

19. *Traveling:* You are planning a trip of 615 miles (round trip), and you know that your car gets an average of $27\frac{1}{3}$ miles per gallon of gas. You also know that your gas tank holds $15\frac{1}{2}$ gallons of gas.

    a. How many gallons of gas will you use on this trip?

    b. If the gas you buy costs $2 per gallon, how much should you plan to spend on this trip for gas?

Writing & Thinking

20. If two fractions are between 0 and 1, can their product be more than 1? Explain.

21. Explain the process of multiplying two fractions. Give an example of a product that cannot be reduced.

22. Explain why the number 0 has no reciprocal.

23. Is division a commutative operation? Explain briefly and give three examples using fractions to help justify your answer.

24. Suppose the product of $5\frac{7}{10}$ and some other number is $10\frac{1}{2}$. Answer the following questions without doing any calculations. Do you think that this other number is more than 1 or less than 1? Why?

7.R.2 Least Common Multiple (LCM)

The **multiples** of a number are _____

Least Common Multiple (LCM)

The **least common multiple (LCM)** of two (or more) counting numbers is _____

_____

DEFINITION

To Find the LCM of a Set of Counting Numbers

1. Find the _____

2. List the _____

3. Find the product of these primes using each _____

_____

PROCEDURE

▶ Watch and Work

Watch the video for Example 4 in the software and follow along in the space provided.

Example 4 Finding the Least Common Multiple (LCM)

Find the LCM of 27, 30, and 42.

Solution

✏ Now You Try It!

Use the space provided to work out the solution to the next example.

Example A Finding the Least Common Multiple (LCM)

Find the LCM of 36, 45, and 60.

Finding Equivalent Fractions

To find a fraction equivalent to $\frac{a}{b}$, multiply the _____

_____

$$\frac{a}{b} = \frac{a}{b}.$$ _____

For example, _____

PROCEDURE

7.R.2 Exercises

Concept Check

True/False. Determine whether each statement is true or false. If a statement is false, explain how it can be changed so the statement will be true. (**Note:** There may be more than one acceptable change.)

1. The LCM of 15 and 25 is 50.

2. The first five multiples of 9 are 9, 18, 27, 36, and 45.

3. The first five multiples of 4 are 4, 8, 12, 20, and 24.

4. When given larger numbers, the most efficient way to find the LCM is to use the prime factorization method.

Practice

Find the LCM of each set of numbers.

5. 6, 10

6. 3, 4, 8

7. **For** 14, 35, and 49, **a.** find the LCM and **b.** state how many times each number divides into the LCM.

For each equation, find the missing numerator that will make the fractions equivalent.

8. $\dfrac{5}{8} = \dfrac{?}{24}$

9. $\dfrac{5}{12} = \dfrac{?}{108}$

Applications

Solve.

10. *Security:* Three security guards meet at the front gate for coffee before they walk around inspecting buildings at a manufacturing plant. The guards take 15, 20, and 30 minutes, respectively, for the inspection trip.

    a. If they start at the same time, in how many minutes will they meet again at the front gate for coffee?

    b. How many trips will each guard have made?

11. *Fruit:* A fruit production company has three packaging facilities, each of which uses different-sized boxes as follows: 24 pieces/box, 36 pieces/box, and 45 pieces/box.

    a. Assuming that the truck provides the same quantity of uniformly-sized pieces of fruit to all three packaging facilities, what is the minimum number of pieces of fruit that will be delivered so that no fruit will be left over?

    b. How many boxes will each facility package?

Writing & Thinking

12. Explain, in your own words, why each number in a set divides evenly into the LCM of that set of numbers.

13. Explain why simply multiplying two numbers together will not necessarily find the LCM of those numbers. Give an example of when it would find the LCM and an example when it would not.

7.R.3 **Addition and Subtraction with Fractions**

To Add Fractions with the Same Denominator

1. _____

2. _____ $\dfrac{a}{b} + \dfrac{c}{b} =$ _____

3. _____

For example, _____

To Add Fractions with Different Denominators

1. Find the _____

2. Change each fraction into _____

3. _____

4. _____

Commutative Property of Addition

The order of the fractions being added can be _____

$$\dfrac{a}{b} + \dfrac{c}{d} = \text{_____}$$

For example, _____

Associative Property of Addition

The grouping of the fractions being added can be _____

$$\dfrac{a}{b} + \left(\dfrac{c}{d} + \dfrac{e}{f} \right) = \text{_____}$$

For example, _____

To Subtract Fractions with the Same Denominator

1. _____

2. _____ $\dfrac{a}{b} - \dfrac{c}{b} =$ _____

3. _____

For example,

_____

PROCEDURE

To Subtract Fractions with Different Denominators

1. Find the _____

2. Change each fraction into _____

3. _____

4. _____

PROCEDURE

▶ Watch and Work

Watch the video for Example 11 in the software and follow along in the space provided.

Example 11   Subtracting Fractions with Different Denominators

Subtract: $1 - \dfrac{5}{8}$

Solution

✏ **Now You Try It!**

Use the space provided to work out the solution to the next example.

Example A Subtracting Fractions with Different Denominators

Subtract: $3 - \dfrac{5}{12}$

7.R.3 **Exercises**

Concept Check

True/False. Determine whether each statement is true or false. If a statement is false, explain how it can be changed so the statement will be true. (**Note:** There may be more than one acceptable change.)

1. The final step in adding fractions is to reduce, if possible.

2. The process for finding the LCD is the same as the process for finding the LCM.

3. LCD represents the Least Common Digit.

4. When subtracting fractions, simply subtract the numerators and the denominators.

5. Subtraction of fractions requires that the fractions have the same denominators.

Practice

Add and reduce to lowest terms.

6. $\dfrac{3}{25} + \dfrac{12}{25}$

7. $\dfrac{2}{7} + \dfrac{4}{21} + \dfrac{1}{3}$

Subtract and reduce to lowest terms.

8. $\dfrac{7}{8} - \dfrac{5}{8}$

10. $2 - \dfrac{9}{16}$

9. $\dfrac{9}{14} - \dfrac{2}{21}$

Applications

Solve.

11. **Postage:** Three pieces of mail weigh $\frac{1}{2}$ ounce, $\frac{1}{5}$ ounce, and $\frac{3}{10}$ ounce. What is the total weight of the letters?

12. **Cooking:** A recipe calls for the following spices: $\frac{1}{2}$ teaspoon of turmeric, $\frac{1}{4}$ teaspoon of ginger, and $\frac{1}{8}$ teaspoon of cumin. What is the total quantity of these three spices?

Writing & Thinking

13. Explain how finding the LCM relates to LCDs.

14. Give an example of a situation where you might add or subtract fractions (other than in class).

7.R.4 Decimal Numbers and Fractions

To Change from Decimal Numbers to Fractions

A decimal number less than 1 (digits are to the right of the decimal point) can be written in fraction form by writing a fraction with:

1. A numerator that consists of the whole number formed by _____

   _____

2. A denominator that is the power of _____

   (For example, a denominator of 100 corresponds to _____.)

   PROCEDURE

1. _____ numbers can be **repeating** or **nonrepeating**. A **repeating decimal**

   has a repeating pattern to its digits.

2. Every fraction with a whole number numerator and nonzero denominator is _____

   _____. Such numbers are called _____.

▶ Watch and Work

Watch the video for Example 8 in the software and follow along in the space provided.

Example 8 Simplifying Expressions with Decimals and Fractions

Find the sum $10\dfrac{1}{2} + 7.32 + 5\dfrac{3}{5}$ in decimal form.

Solution

✏ Now You Try It!

Use the space provided to work out the solution to the next example.

Example A Simplifying Expressions with Decimals and Fractions

Find the sum $2.88 + \dfrac{1}{4} + 13\dfrac{9}{10}$ in decimal form.

7.R.4 Exercises

Concept Check

True/False. Determine whether each statement is true or false. If a statement is false, explain how it can be changed so the statement will be true. (**Note:** There may be more than one acceptable change.)

1. When a decimal number is changed to a fraction, the denominator will be the power of 10 that names the rightmost digit of the decimal number.

2. When a decimal number is changed to a fraction, the numerator can be determined by using the whole number that is formed by all the digits of the decimal number.

3. Fractions can always be converted to decimal form without losing accuracy.

4. In decimal form, $\dfrac{1}{3}$ is repeating and nonterminating.

Practice

Change each decimal number to a fraction or mixed number in lowest terms.

5. 0.18

6. 2.75

Change each fraction to a decimal number rounded to the nearest hundredth.

7. $\dfrac{20}{3}$

8. $\dfrac{40}{9}$

Simplify the expression by first writing all of the numbers in decimal form. Round to the nearest hundredth, if necessary.

9. $\dfrac{1}{4} + 0.25 + \dfrac{1}{5}$

10. Arrange $0.76, \dfrac{3}{4}, \dfrac{7}{10}$ in order from smallest to largest.

Applications

Solve.

11. *Rectangles:* A rectangle measures 6.4 inches in length, and has a width that measures $\frac{2}{5}$ of the length. Find the perimeter of the rectangle.

12. *Groceries:* A loaf of bread weighs 21.6 ounces. Mauricio cut off a third of the loaf to save for later and then cut the remaining portion into 16 equal slices. What was the weight of each slice of the 16 slices he cut?

Writing & Thinking

13. Describe the process used to change a terminating decimal number to a fraction.

14. List 2 different ways to solve this problem: $\frac{1}{2} + 3.67 - \frac{1}{8}$. State which method you prefer and why.

Review Concepts

for *Statistics*

8.R.1 Decimals and Percents

The word percent comes from the Latin *per centum*, meaning _____. So **percent means** _____, or **the ratio of a number to** _____.

The symbol % is called the _____. This sign has the same meaning as the fraction $\frac{1}{100}$

To Change a Decimal Number to a Percent

1. Move the _____

2. Write the _____

PROCEDURE

To Change a Percent to a Decimal Number

1. Move the _____

2. Delete the _____

PROCEDURE

▶ Watch and Work

Watch the video for Example 3 in the software and follow along in the space provided.

Example 3 Changing Percents to Decimal Numbers

Change each percent to a decimal number.

a. 76.%

b. 18.5%

c. 50%

d. 100%

e. 0.25%

✏ **Now You Try It!**

Use the space provided to work out the solution to the next example.

Example A Changing Percents to Decimal Numbers

Change each percent to a decimal number.

a. 40%

b. 211%

c. 0.6%

d. 29.37%

e. 102%

8.R.1 **Exercises**

Concept Check

True/False. Determine whether each statement is true or false. If a statement is false, explain how it can be changed so the statement will be true. (**Note:** There may be more than one acceptable change.)

1. If a decimal number is less than 1, then the equivalent percent will be less than 100%.

2. It is not possible to have a percent greater than 100%.

3. A decimal number that is between 0.01 and 0.10 is between 10% and 100%.

4. To change from a percent to a decimal, simply omit the percent sign.

Practice

Change each fraction to a percent.

5. $\dfrac{20}{100}$

6. $\dfrac{125}{100}$

Change each decimal number to a percent.

7. 0.02

8. 2.3

Change each percent to a decimal number.

9. 7%

10. 179%

Applications

Solve.

11. *Interest:* A savings account is offering an interest rate of 0.04 for the first year after opening the account. Change 0.04 to a percent.

12. *Sales Tax:* Suppose that sales tax is figured at 7.25%. Change 7.25% to a decimal.

Writing & Thinking

13. Describe the relationship between percent and the number 100.

14. Describe a situation where more than 100% is possible. Describe a situation where it is impossible to have more than 100%.

8.R.2 Fractions and Percents

> ## To Change a Fraction to a Percent
>
> 1. Change the _____. (Divide _____.)
>
> 2. Change the _____.
>
> **PROCEDURE**

▶ Watch and Work

Watch the video for Example 3 in the software and follow along in the space provided.

Example 3 Changing Mixed Numbers to Percents

Change $2\frac{1}{4}$ to a percent.

Solution

✏ Now You Try It!

Use the space provided to work out the solution to the next example.

Example A Changing Mixed Numbers to Percents

Change $1\frac{1}{2}$ to a percent.

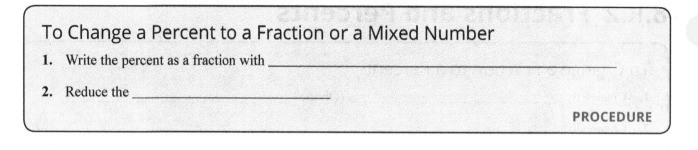

> ## To Change a Percent to a Fraction or a Mixed Number
>
> 1. Write the percent as a fraction with _____
>
> 2. Reduce the _____
>
> PROCEDURE

8.R.2 Exercises

Concept Check

True/False. Determine whether each statement is true or false. If a statement is false, explain how it can be changed so the statement will be true. (**Note:** There may be more than one acceptable change.)

1. Fractions that have denominators other than 100 cannot be changed to a percent.

2. The fraction $\dfrac{1}{5}$ is equivalent to $\dfrac{1}{5}\%$.

3. When changing from a percent to a mixed number, the fraction does not need to be reduced.

4. $75\% = 0.75 = \dfrac{3}{4}$

Practice

Change each fraction or mixed number to a percent. If necessary, round to the nearest tenth of a percent.

5. $\dfrac{3}{4}$

6. $5\dfrac{3}{10}$

Change each percent to a fraction or mixed number and reduce, if possible.

7. 120%

8. 12.5%

Applications

Solve.

9. *Exam Grades:* Out of a possible total of 240 points on an exam, David received 204 points. What percent of the exam did David get correct?

10. *College Degrees:* To receive a Bachelor of Science (BS) degree at Bluefield State College, the student must complete a total of 128 credit hours, of which 41 of these credits must be general education Core Skills courses. What percent of the total curriculum is dedicated to general education courses? [1]

---

1 Source: 2010–2011 Bluefield, WV State College Catalogue, p.76

Writing & Thinking

11. Justify why mixed numbers are a larger percentage than proper fractions alone. (Consider the value of 100%.)

12. Describe the process to change a percent to a fraction or mixed number.

8.R.3 Working with Formulas

1. **Formulas** are general rules or _____ stated _____.

2. We say that the formula $d = rt$ is _____ d _____ r and t. Similarly,

   the formula $A = \frac{1}{2}bh$ is solved for _____ in terms of _____, and the formula

   $P = R - C$ (profit is equal to revenue minus cost) is solved for _____ in terms of _____

   _____.

▶ Watch and Work

Watch the video for Example 6 in the software and follow along in the space provided.

Example 6 Solving for Different Variables

Given $V = \dfrac{k}{P}$, solve for P in terms of V and k.

Solution

✏ Now You Try It!

Use the space provided to work out the solution to the next example.

Example A Solving for Different Variables

Given $P = \dfrac{I}{rt}$ solve for t in terms of I, r, and P.

8.R.3 Exercises

Concept Check

True/False. Determine whether each statement is true or false. If a statement is false, explain how it can be changed so the statement will be true. (**Note:** There may be more than one acceptable change.)

1. When using formulas, typically it does not matter if capital or lower case letters are used: $A = a$, $C = c$, etc.

2. If the perimeter and length are known, $P = 2l + 2w$ can be used to find the width of a rectangle.

3. Rate of interest is stated as an annual rate in percent form.

Applications

In the following application problems, read the descriptions carefully and then substitute the values given in the problem for the corresponding variables in the formulas. Evaluate the resulting expression for the unknown variable.

Velocity

If an object is shot upward with an initial velocity v_0 in feet per second, the velocity v in feet per second is given by the formula $v = v_0 - 32t$, where t is time in seconds. (v_0 is read "v sub zero." The $_0$ is called a subscript.)

4. An object projected upward with an initial velocity of 106 feet per second has a velocity of 42 feet per second. How many seconds have passed?

Investments

The total amount of money in an account with P dollars invested in it is given by the formula $A = P + Prt$, where r is the rate expressed as a decimal and t is time (one year or part of a year).

5. If $1000 is invested at 6% interest, find the total amount in the account after 6 months.

Cost

The total cost C of producing x items can be found by the formula $C = ax + k$, where a is the cost per item and k is the fixed costs (rent, utilities, and so on).

6. Find the total cost of producing 30 items if each costs $15 and the fixed costs are $580.

Solve each formula for the indicated variable.

7. $P = a + b + c$; solve for b.

8. $P = 3s$; solve for s.

9. $I = Prt$; solve for t.

10. $A = P(1 + rt)$; solve for r.

8.R.4 The Cartesian Coordinate System

1. Descartes based his system on a relationship between _____ in a plane and _____

   of real numbers.

2. In the ordered pair (x, y), x is called the _____ and y is called the _____

   _____.

3. In an ordered pair of the form (x, y), the _____ is called the **independent**

   variable and the _____ is called the **dependent variable**.

4. The Cartesian coordinate system relates algebraic equations and ordered pairs to geometry. In this system,

   two number lines intersect at right angles and separate the plane into four _____. The **origin**,

   designated by the ordered pair $(0, 0)$, is _____. The

   horizontal number line is called the _____ or _____. The vertical number line

   is called the _____ or _____.

One-to-One Correspondence

_____

_____

DEFINITION

▶ Watch and Work

Watch the video for Example 4 in the software and follow along in the space provided.

Example 4 Finding Ordered Pairs

Complete the table so that each ordered pair will satisfy the equation $y = -3x + 1$.

x	y	(x, y)
0		
	4	
$\dfrac{1}{3}$		
3		

Solution

✎ Now You Try It!

Use the space provided to work out the solution to the next example.

Example A Finding Ordered Pairs

Complete the table so that each ordered pair will satisfy the equation $y = -3x + 2$.

x	y	(x, y)
0		
	1	
−2		
	0	

8.R.4 **Exercises**

True/False. Determine whether each statement is true or false. If a statement is false, explain how it can be changed so the statement will be true. (**Note:** There may be more than one acceptable change.)

1. The graph of every ordered pair that has a positive x-coordinate and a negative y-coordinate can be found in Quadrant IV.

2. To find the y-value that corresponds with $x = 2$, substitute 2 for x into the given equation and solve for y.

3. If $(-7, 3)$ is a solution of $y = 3x + 24$, then $(-7, 3)$ satisfies $y = 3x + 24$.

4. If point $A = (0, 4)$, then point A lies on the x-axis.

Practice

List the set of ordered pairs corresponding to the points on the graph.

5.

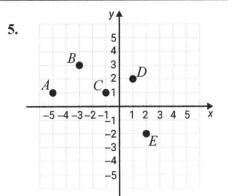

Plot each set of ordered pairs and label the points.

6. $\{A(4,-1), B(3,2), C(0,5), D(1,-1), E(1,4)\}$

Determine the missing coordinate in each of the ordered pairs so that the point will satisfy the equation given.

7. $x - 2y = 2$

   a. $(0, \underline{\quad})$

   b. $(4, \underline{\quad})$

   c. $(\underline{\quad}, 0)$

   d. $(\underline{\quad}, 3)$

Complete the tables so that each ordered pair will satisfy the given equation. Plot the resulting sets of ordered pairs.

8. $y = 2x - 3$

x	y
0	
	-1
-2	
	3

Determine which, if any, of the ordered pairs satisfy the given equation.

9. $2x - 3y = 7$

   a. $(1, 3)$

   b. $\left(\dfrac{1}{2}, -2\right)$

   c. $\left(\dfrac{7}{2}, 0\right)$

   d. $(2, 1)$

The graph of a line is shown. List any three points on the line. (There is more than one correct answer.)

10.

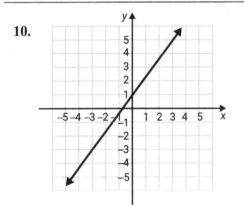

Applications

Solve.

11. ***Exchange Rates:*** At one point in 2017, the exchange rate from US dollars to Euros was $E = 0.85D$ where E is Euros and D is dollars.

   a. Make a table of ordered pairs for the values of D and E if D has the values $100, $200, $300, $400, and $500.

   b. Plot the points corresponding to the ordered pairs.

12. ***Temperature:*** Given the equation $F = \frac{9}{5}C + 32$ where C is temperature in degrees Celsius and F is the corresponding temperature in degrees Fahrenheit:

   a. Make a table of ordered pairs for the values of C and F if C has the values $-20°$, $-10°$, $-5°$, $0°$, $5°$, $10°$, and $15°$.

   b. Plot the points corresponding to the ordered pairs.

8.R.5 Graphing Linear Equations in Two Variables

Standard Form of a Linear Equation

Any equation of the form

_____

where A, B, and C are real numbers and A and B are not both equal to 0, is called the standard form of a linear equation.

DEFINITION

To Graph a Linear Equation in Two Variables

1. Locate any two points that _____

2. _____

3. _____

4. To check: Locate a third point that _____

_____

PROCEDURE

▶ Watch and Work

Watch the video for Example 2 in the software and follow along in the space provided.

Example 2 Graphing a Linear Equation in Two Variables

Graph: $2x + 3y = 6$

Solution

● **Now You Try It!**

Use the space provided to work out the solution to the next example.

Example A Graphing a Linear Equation in Two Variables

Graph: $3x + 2y = 6$

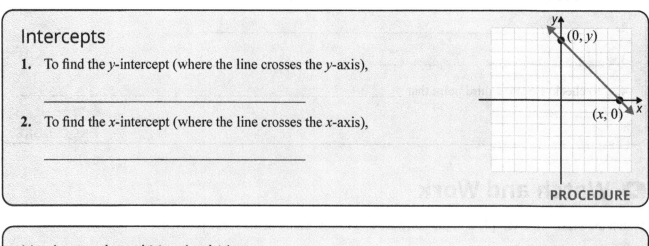

Intercepts

1. To find the y-intercept (where the line crosses the y-axis),

   _____

2. To find the x-intercept (where the line crosses the x-axis),

   _____

PROCEDURE

Horizontal and Vertical Lines

For real numbers a and b, the graph of _____

DEFINITION

8.R.5 Exercises

Concept Check

True/False. Determine whether each statement is true or false. If a statement is false, explain how it can be changed so the statement will be true. (**Note:** There may be more than one acceptable change.)

1. The y-intercept is the point where a line crosses the y-axis.

2. The terms ordered pair and point are used interchangeably.

3. A horizontal line does not have a y-intercept.

4. All x-intercepts correspond to an ordered pair of the form $(0, y)$.

Practice

Graph each linear equation by locating at least two ordered pairs that satisfy the given equation.

5. $x + y = 3$ **7.** $y = -3$

6. $x = 1$

Graph each linear equation by locating the x-intercept and the y-intercept.

8. $y = 4x - 10$ **9.** $3x - 7y = -21$

Applications

Solve.

10. *Chemistry:* The amount of potassium in a clear bottle of a popular sports drink declines over time when exposed to the UV lights found in most grocery stores. The amount of potassium in a container of this sports drink is given by the equation $y = -30x + 360$, where y represents the mg of potassium remaining after x days on the shelf. Find both the x-intercept and y-intercept, and interpret the meaning of each in the context of this problem.

11. *Education:* Mr. Adler has found that the grade each student gets in his Introductory Algebra course directly correlates with the amount of time spent doing homework, and is represented by the equation $y = 7x + 30$, where y represents the numerical score the student receives on an exam (out of 100 points) after spending x hours per week doing homework. Find the y-intercept and interpret its meaning in this context.

Writing & Thinking

12. Explain, in your own words, why it is sufficient to find the x-intercept and y-intercept to graph a line (assuming that they are not the same point).

13. Explain, in your own words, how you can determine if an ordered pair is a solution to an equation.

8.R.6 Slope-Intercept Form

1. For a line, the _____ is called the **slope of the line**.

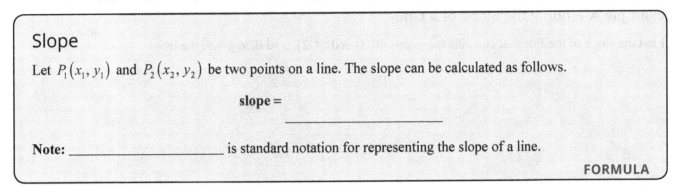

Slope

Let $P_1(x_1, y_1)$ and $P_2(x_2, y_2)$ be two points on a line. The slope can be calculated as follows.

$$\text{slope} = \underline{\hspace{3cm}}$$

Note: _____ is standard notation for representing the slope of a line.

FORMULA

▶ Watch and Work

Watch the video for Example 2 in the software and follow along in the space provided.

Example 2 Finding the Slope of a Line

Find the slope of the line that contains the points $(1, 3)$ and $(5, 1)$, and then graph the line.

Solution

✏️ Now You Try It!

Use the space provided to work out the solution to the next example.

Example A Finding the Slope of a Line

Find the slope of the line that contains the points $(0, 5)$ and $(4, 2)$, and then graph the line.

Positive and Negative Slope

Lines with positive slope go _____

Lines with negative slope go _____

DEFINITION

Horizontal and Vertical Lines

The following two general statements are true for horizontal and vertical lines.

1. For horizontal lines (of the form _____

2. For vertical lines (of the form _____

DEFINITION

Slope-Intercept Form

_____ is called the slope-intercept form for the equation of a line, where m is the slope and $(0, b)$ is the y-intercept.

DEFINITION

8.R.6 Exercises

Practice

Find the slope of the line determined by each pair of points.

1. $(1, -2); (1, 4)$

2. $(-3, 7); (4, -1)$

Determine whether the equation $x = -3$ represents a horizontal line or a vertical line and give its slope.

3. $x = -3$

Write each equation in slope-intercept form. Find the slope and y-intercept, and then use them to draw the graph.

4. $y = 2x - 1$

5. $3y - 9 = 0$

Find an equation in slope-intercept form for the line passing through (0,3) with the slope $m = -\dfrac{1}{2}$.

6. $(0, 3); m = -\dfrac{1}{2}$

Applications

Solve.

7. ***Purchases:*** John bought his new car for $35,000 in the year 2014. He knows that the value of his car has depreciated linearly. If the value of the car in 2017 was $23,000, what was the annual rate of depreciation of his car? Show this information on a graph. (When graphing, use years as the x-coordinates and the corresponding values of the car as the y-coordinates.)

8. ***Cell Phones:*** The number of people in the United States with mobile cellular phones was about 198 million in 2011 and about 232 million in 2016. If the growth in the usage of mobile cellular phones was linear, what was the approximate rate of growth per year from 2011 to 2016. Show this information on a graph. (When graphing, use years as the x-coordinates and the corresponding numbers of users as the y-coordinates.)[1]

Writing & Thinking

9. **a.** Explain in your own words why the slope of a horizontal line must be 0.

   b. Explain in your own words why the slope of a vertical line must be undefined.

8.R.7 Evaluating Radicals

> ## Radical Terminology
>
> The symbol $\sqrt{}$ is called _____
>
> The number under the _____
>
> The complete expression, such as $\sqrt{64}$ is called _____
>
> **DEFINITION**

> ## Square Root
>
> If a is a nonnegative real number, then
>
> _____
>
> _____
>
> _____
>
> **DEFINITION**

> ## Cube Root
>
> If a is a real number, then _____
>
> **DEFINITION**

▶ Watch and Work

Watch the video for Example 4 in the software and follow along in the space provided.

Example 4 Evaluating Cube Roots

a. Because $2^3 =$ _, $\sqrt[3]{8} =$ _.

b. Because $(-6)^3 =$ _____, $\sqrt[3]{-216} =$ _.

_____

c. Because $\left(\dfrac{1}{3}\right)^3 =$ __, $\sqrt[3]{\dfrac{1}{27}} =$ _.

✏️ Now You Try It!

Use the space provided to work out the solution to the next example.

Example A Evaluating Cube Roots

Evaluate the following radical expressions.

    a. $\sqrt[3]{64}$

    b. $\sqrt[3]{-125}$

    c. $\sqrt[3]{\dfrac{1}{1000}}$

8.R.7 Exercises

Concept Check

True/False. Determine whether each statement is true or false. If a statement is false, explain how it can be changed so the statement will be true. (**Note:** There may be more than one acceptable change.)

1. If a number is squared and the principal square root of the result is found, that square root is always equal to the original number.

2. There is no real number that can be a square root of a negative number.

3. The index is the number underneath the radical sign.

4. The cube root of −27 is a real number.

Practice
Simplify the following square roots and cube roots.

5. $\sqrt{49}$

6. $\sqrt{289}$

7. $\sqrt[3]{1000}$

8. $\sqrt[3]{\dfrac{27}{64}}$

9. $\sqrt{0.04}$

Applications
Solve.

10. *Area:* The area of a square tile is 16 square inches.

    a. How long are the sides of the square tile?

    b. How many tiles would be needed for a four-foot-long and four-inch-high backsplash in a newly designed bathroom?

11. *Volume:* The volume of a child's building block is 64 cubic centimeters.

    a. Assuming the building block is a perfect cube, find the length of each side of the block.

    b. If a child stacks 5 blocks directly on top of each other, find the height of the structure that is created.

Writing & Thinking

12. Discuss, in your own words, why the square root of a negative number is not a real number.

13. Discuss, in your own words, why the cube root of a negative number is a negative number.

Review Concepts

for *Personal Finance*

9.R.1 Introduction to Whole Numbers

Whole Numbers

The **whole numbers** are _____ along with _____

Natural numbers = $\mathbb{N}$ = {_____}

Whole numbers = $\mathbb{W}$ = {_____}

DEFINITION

The Decimal System

The **decimal system** (or base ten system) is a place value system that depends on three things.

1. _____

2. the placement of _____

3. the value of _____

DEFINITION

1. In **expanded notation**, the values represented by each digit are _____

▶ Watch and Work

Watch the video for Example 3 in the software and follow along in the space provided.

Example 3 Writing Numbers in Expanded Notation

Write each number in expanded notation.

a. 954

b. 6507

Solution

✏ Now You Try It!

Use the space provided to work out the solution to the next example.

Example A Application: Finding the Mean

Write each number in expanded notation.

a. 463

b. 7300

Reading and Writing Whole Numbers

You should note the following four things when reading or writing whole numbers.

1. Digits are read in _____ (groups _____).

2. Commas are used to _____ if a number has _____

3. The word "and" does not _____

_____

4. Hyphens are used to write words for _____

9.R.1 Exercises

Concept Check

True/False. Determine whether each statement is true or false. If a statement is false, explain how it can be changed so the statement will be true. (**Note:** There may be more than one acceptable change.)

1. In the number 21,057, the "1" represents 1000.

2. 56,317 can be written as $56,000 + 300 + 17$ in expanded notation.

3. 42,360 can be written as forty-two thousand, three hundred sixty.

4. The word "and" is not used when reading or writing whole numbers.

Practice

5. Name the place value of each nonzero digit in the following number: 24,608.

6. Write 1892 in expanded notation.

7. Write 683,100 in words.

Write each number in standard notation.

8. Write four hundred thousand, seven hundred thirty-six in standard notation.

Applications

9. *Lakes:* The largest lake in the United States is Lake Superior. It takes up an area of 82,103 square kilometers. Write 82,103 in words.

10. *Card Games:* The largest collection of Joker playing cards consists of eight thousand, five hundred twenty cards amassed by Tony De Santis after inheriting a two thousand piece collection from the magician Fernando Riccardi. Write eight thousand, five hundred twenty in standard form.

Writing & Thinking

11. How are natural numbers and whole numbers different and how are they the same?

12. When are hyphens used to write numbers in English words?

9.R.2 Addition and Subtraction with Whole Numbers

Adding Whole Numbers

1. Write the numbers _____ so that the _____

2. Add only the _____

Carrying When Adding Whole Numbers

If the sum of the digits in one column is more than 9,

1. write the _____

2. carry the _____ as a number to be added to _____

   _____

Commutative Property of Addition

The order of the numbers in addition can be _____

For example, _____

Associative Property of Addition

The grouping of the numbers in addition can be _____

For example, _____

Additive Identity Property

The sum of a number and 0 is _____

For example, _____

The number 0 is called the _____

1. The **perimeter** of a geometric figure is _____

▶ Watch and Work

Watch the video for Example 6 in the software and follow along in the space provided.

Example 6 Calculating the Perimeter of a Polygon

Calculate the perimeter of the polygon.

(**Note:** A 5-sided polygon is called a **pentagon**.)

Solution

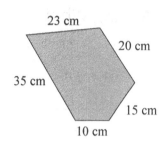

✎ Now You Try It!

Use the space provided to work out the solution to the next example.

Example A Calculating the Perimeter of a Polygon

Calculate the perimeter of the polygon.

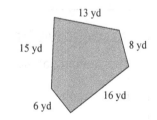

Subtraction

Subtraction is the operation of _____

The **difference** is the result of _____

_____

DEFINITION

Subtracting Whole Numbers

1. Write the numbers _____ so that the _____

2. Subtract only the _____

3. Check by _____ The sum must be _____

PROCEDURE

9.R.2 **Exercises**

Concept Check

True/False. Determine whether each statement is true or false. If a statement is false, explain how it can be changed so the statement will be true. (**Note:** There may be more than one acceptable change.)

1. A polygon is a geometric figure in a plane with two or more sides.

2. To find the perimeter of a rectangle, add the lengths of the four sides.

3. When subtracting, sometimes the digit being subtracted is larger than the digit it is being subtracted from and so "carrying" must occur.

4. If your bank account has a balance of $743 and you want to withdraw $115, you would use subtraction to find that the new balance would be $628.

Practice

Simplify.

5. $\begin{array}{r} 15 \\ +43 \\ \hline \end{array}$

6. $\begin{array}{r} 981 \\ +46 \\ \hline \end{array}$

7.   275
     −131

8.   543
     −167

9. Calculate the perimeter of the given geometric figure.

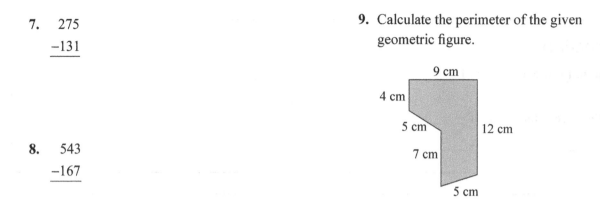

Applications

Solve.

10. **Budgeting:** The Magley family has the following monthly budget: $815 mortgage; $69 electric; $47 water; and $122 phone bills (including cell phones). What is the family's budget for each month for these expenses?

11. **Credit:** A couple sold their house for $135,000. They paid the realtor $8100, and other expenses of the sale came to $800. If they owed the bank $87,000 for the mortgage, what were their net proceeds from the sale?

Writing & Thinking

12. Explain when "carrying" should be used in addition with whole numbers and give an example.

13. Explain when "borrowing" would be used in subtraction and give an example.

9.R.3 Exponents and Order of Operations

1. When looking at $3^5 = 243$, 3 is the _____, 5 is the _____, and 243

   is the _____. _____ are written slightly to the right and above

   the _____. The expression 3^5 is _____.

The Exponent 1

Any number raised to the first power _____

For example, _____

DEFINITION

The Exponent 0

Any nonzero number raised to the 0 power _____

For example, _____

Note: The expression 0^0 _____

DEFINITION

Rules for Order of Operations

1. Simplify within grouping symbols, such as _____
   (If there are more than one pair of grouping symbols, start with _____
   _____)

2. Evaluate any _____

3. Moving from left to right, perform any _____
   _____

4. Moving from left to right, perform any _____
   _____

PROCEDURE

---

2. A well-known mnemonic device for remembering the rules for order of operations is the following.

Please	Excuse	My	Dear	Aunt	Sally
↓	↓	↓	↓	↓	↓
_____	_____	_____	_____	_____	_____

▶ Watch and Work

Watch the video for Example 6 in the software and follow along in the space provided.

Example 6 Using the Order of Operations with Whole Numbers

Simplify: $2 \cdot 3^2 + 18 \div 3^2$

Solution

✏ Now You Try It!

Use the space provided to work out the solution to the next example.

Example A Using the Order of Operations with Whole Numbers

Simplify: $6^2 \div 9 + 3 - 14 \div 7$

9.R.3 **Exercises**

Concept Check

True/False. Determine whether each statement is true or false. If a statement is false, explain how it can be changed so the statement will be true. (**Note:** There may be more than one acceptable change.)

1. Nine squared is equal to eighteen.

2. $2^7 = 128$

3. 7^0 is undefined.

4. According to the order of operations, multiplication is always performed before division.

Practice

For each exponential expression **a.** identify the base, **b.** identify the exponent, and **c.** evaluate the exponential expression.

5. 2^3

6. 4^0

Simplify.

7. $18 \div 2 - 1 - 3 \cdot 2$

8. $30 \div 2 - 11 + 2(5-1)^3$

Applications

Solve.

9. *Card Games:* Neville bought 15 boxes of trading cards. Each box has 10 packs of trading cards. Each pack of trading cards contains 20 cards. He adds 132 cards that he already owns to the newly purchased cards. Then, Neville evenly distributes all of the cards to 6 of his friends. How many trading cards would each person get?

   a. If you simplify the expression $15 \cdot 10 \cdot 20 + 132 \div 6$ using the order of operations, will you get the correct answer? If not, explain what is wrong with the expression.

   b. What is the answer? If necessary, write the corrected expression to get the correct results when following the order of operations.

10. *Purchases:* Robert is purchasing shirts for his weekend soccer team. The shirts he wants to buy are normally $25 each but are on sale for $10 off. His team has a total of 11 players. How much will he spend to buy the shirts?

    a. If you simplify the expression $\$25 - \$10 \cdot 11$ using the order of operations, will you get the correct answer? If not, explain what is wrong with the expression.

    b. What is the answer? If necessary, write the corrected expression to get the correct results when following the order of operations.

Writing & Thinking

11. Give one example where addition should be completed before multiplication.

9.R.4 Introduction to Decimal Numbers

To Read or Write a Decimal Number

1. Read (or write) the _____

2. Read (or write) the _____

3. Read (or write) the _____.

   Then, name the fraction part with the _____.

PROCEDURE

To Compare Two Decimal Numbers

1. Moving left to right, compare digits _____

2. When one compared digit is larger, then the _____

PROCEDURE

▶ Watch and Work

Watch the video for Example 6 in the software and follow along in the space provided.

Example 6 Comparing Decimal Numbers

Arrange the following three decimal numbers in order from smallest to largest: 6.67, 5.14, 6.28. Then, graph them on a number line.

Solution

✏ Now You Try It!

Use the space provided to work out the solution to the next example.

Example A Comparing Decimal Numbers

Arrange the following three decimal numbers in order from smallest to largest: 2.6, 2.06, 2.46. Then, graph them on a number line.

Rounding Rule for Decimal Numbers

1. Look at the single digit one place value to the right of the digit in the place of desired accuracy.

   a. **If this digit is less than 5,** _____

      and replace all digits to the right with zeros. All digits to the _____.

   b. **If this digit is 5 or greater,** _____

      and replace all digits to the right with zeros. All digits to the _____

      _____. Then, the 9 is replaced by 0 and the _____

      _____

2. Zeros to the right of the place of accuracy that are also to the right of the _____

   _____

   PROCEDURE

9.R.4 **Exercises**

Concept Check

True/False. Determine whether each statement is true or false. If a statement is false, explain how it can be changed so the statement will be true. (**Note:** There may be more than one acceptable change.)

  1. Two hundred thousand, four hundred six and twelve hundredths can be written as 200,406.12.

2. 92.586 is greater than 92.6.

3. On a number line, any number to the left of another number is larger than that other number.

4. When a decimal number is rounded, all numbers to the right of the place of accuracy become zeros in the final answer.

Practice

5. Write $2\dfrac{57}{100}$ in decimal notation.

6. Write 20.7 in words.

7. Write six and twenty-eight thousandths in decimal notation.

8. Arrange 0.2, 0.26, and 0.17 in order from smallest to largest. Then, graph the numbers on a number line.

Fill in the blanks to correctly complete each statement.

9. Round 3.00652 to the nearest ten-thousandth.

    a. The digit in the ten-thousandths position is ___ .

    b. The next digit to the right is ___ .

    c. Since ___ is less than 5, leave ___ as it is and replace ___ with 0.

    d. So 3.00652 rounds to _____ to the nearest ten-thousandth.

Applications

In each exercise, write the decimal numbers that are not whole numbers in words.

10. *Unicycles:* The tallest unicycle ever ridden was 114.8 feet tall, and was ridden by Sam Abrahams (with a safety wire suspended from an overhead crane) for a distance of 28 feet in Pontiac, Michigan, on January 29, 2004.[1]

11. *Water Weight:* One quart of water weighs approximately 2.0825 pounds.

Writing & Thinking

12. Discuss situations where you think it is particularly appropriate (or necessary) to write numbers in English word form.

13. With **a.** and **b.** as examples, explain in your own words how you can tell quickly when one decimal number is larger (or smaller) than another decimal number.

    a. The decimal number 2.765274 is larger than the decimal number 2.763895.

    b. The decimal number 17.345678 is larger than the decimal number 17.345578.

1 Source: http://semcycle.biz/record

9.R.5 Decimals and Percents

The word percent comes from the Latin *per centum*, meaning _____. So **percent means** _____, or **the ratio of a number to** _____.

The symbol % is called the _____. This sign has the same meaning as the fraction $\frac{1}{100}$

> ## To Change a Decimal Number to a Percent
>
> 1. Move the _____
>
> 2. Write the _____
>
> **PROCEDURE**

> ## To Change a Percent to a Decimal Number
>
> 1. Move the _____
>
> 2. Delete the _____
>
> **PROCEDURE**

▶ Watch and Work

Watch the video for Example 3 in the software and follow along in the space provided.

Example 3 Changing Percents to Decimal Numbers

Change each percent to a decimal number.

a. 76.%

b. 18.5%

c. 50%

d. 100%

e. 0.25%

✏ **Now You Try It!**

Use the space provided to work out the solution to the next example.

Example A Changing Percents to Decimal Numbers

Change each percent to a decimal number.

a. 40%

b. 211%

c. 0.6%

d. 29.37%

e. 102%

9.R.5 **Exercises**

Concept Check

True/False. Determine whether each statement is true or false. If a statement is false, explain how it can be changed so the statement will be true. (**Note:** There may be more than one acceptable change.)

1. If a decimal number is less than 1, then the equivalent percent will be less than 100%.

2. It is not possible to have a percent greater than 100%.

3. A decimal number that is between 0.01 and 0.10 is between 10% and 100%.

4. To change from a percent to a decimal, simply omit the percent sign.

Practice

Change each fraction to a percent.

5. $\dfrac{20}{100}$

6. $\dfrac{125}{100}$

Change each decimal number to a percent.

7. 0.02

8. 2.3

Change each percent to a decimal number.

9. 7%

10. 179%

Applications

Solve.

11. *Interest:* A savings account is offering an interest rate of 0.04 for the first year after opening the account. Change 0.04 to a percent.

12. *Sales Tax:* Suppose that sales tax is figured at 7.25%. Change 7.25% to a decimal.

Writing & Thinking

13. Describe the relationship between percent and the number 100.

14. Describe a situation where more than 100% is possible. Describe a situation where it is impossible to have more than 100%.

9.R.6 Solving Percent Problems Using Equations

Terms Related to the Basic Equation $R \cdot B = A$

For the basic equation $R \cdot B = A$,

$R =$ _____ (as a _____).

$B =$ _____ (number that we _____).

$A =$ _____ (a part of _____).

Three Basic Types of Percent Problems and the Formula $R \cdot B = A$

Type 1: Find the amount given the _____

For example, what is _____

_____

Type 2: Find the base given the _____

For example, 42% of what _____

_____

Type 3: Find the percent (rate) given the _____

For example, what _____

_____

▶ Watch and Work

Watch the video for Example 4 in the software and follow along in the space provided.

Example 4 Finding the Amount

Find 75% of 56.

Solution

✏ Now You Try It!

Use the space provided to work out the solution to the next example.

Example A Finding the Amount

Find 150% of 60.

9.R.6 **Exercises**

Concept Check

True/False. Determine whether each statement is true or false. If a statement is false, explain how it can be changed so the statement will be true. (**Note:** There may be more than one acceptable change.)

1. In order to solve the equation $0.56 \cdot B = 12$ for the base B one would multiply 12 by 0.56.

2. In the problem "126% of 720 is what number?" the missing number is the amount.

3. The solution to the problem "50% of what number is 352?" could be found by solving the equation $50 \cdot B = 352$.

4. If the base is 120 and the rate is greater than 100%, then the amount will be greater than 120.

Practice

Use the equation $R \cdot B = A$ to find each unknown quantity. Round percents to the nearest tenth of a percent. All other answers should be rounded to the nearest hundredth, if necessary.

5. 10% of 70 is what number?

6. Find 75% of 12.

7. 150% of _____ is 63.

8. What percent of 75 is 15?

9. _____% of 30 is 6.

Applications

Solve.

10. *Presidential Vetoes:* During his presidency, from 1945 to 1953, Harry S. Truman vetoed 250 congressional bills, and 12 of those vetoes were overridden. What percent of Truman's vetoes were overridden?

11. *Mortgages:* The minimum down payment to obtain the best financing rate on a house is 20%. Assuming that John has set aside $35,000 and wants to take advantage of the best financing rate, what is the most expensive house he can purchase? [1]

Writing & Thinking

12. Explain the connection between the proportion $\dfrac{P}{100} = \dfrac{A}{B}$ and the equation $R \cdot B = A$.

13. Explain how to determine which number is the rate (percent), which one is the amount, and which one is the base.

1 Source: http://www.kiplinger.com/magazine/archives/what-it-takes-to-get-a-mortgage.html

9.R.7 Simplifying and Evaluating Algebraic Expressions

Like Terms

Like terms (or similar terms) are terms that are _____

_____

DEFINITION

Combining Like Terms

To combine like terms, _____

DEFINITION

To Evaluate an Algebraic Expression

1. _____

2. _____

3. _____

(**Note:** Terms separated by + and − signs may be evaluated _____

_____.)

PROCEDURE

▶ Watch and Work

Watch the video for Example 5 in the software and follow along in the space provided.

Example 5 Simplifying and Evaluating Algebraic Expressions

Simplify and evaluate $3ab - 4ab + 6a - a$ for $a = 2$, $b = -1$.

Solution

✏ Now You Try It!

Use the space provided to work out the solution to the next example.

Example A Identifying Properties of Addition and Multiplication

Simplify and evaluate

$5ab - 8ab + 2a - 3a$ for $a = -3$, $b = 1$

9.R.7 Exercises

True/False. Determine whether each statement is true or false. If a statement is false, explain how it can be changed so the statement will be true. (**Note:** There may be more than one acceptable change.)

1. A variable that does not appear to have an exponent has an exponent of 1.

2. In the term $-9x$, nine is being subtracted from x.

3. In the term "$12a$," 12 is the constant.

4. Like terms have the same coefficients.

Practice

Identify the like terms in each list of terms.

5. $-5, 3, 7x, 8, 9x, 3y$

Simplify each expression by combining like terms.

6. $8x + 7x$

7. $3x - 5x + 12x$

8. $13x + 12x^2 + 15x - 35 - 41 - 2x^2$

Simplify each expression and then evaluate the expression for $y = 3$ and $a = -2$.

9. $5y + 4 - 2y$

10. $\dfrac{3a + 5a}{-2} + 12a$

Applications

Solve.

11. **Profit:** An apartment management company owns a property with 100 units. The company has determined that the profit made per month from the property can be calculated using the equation $P = -10x^2 + 1500x - 6000$, where x is the number of units rented per month. How much profit does the company make when 80 units are rented?

12. **Physics:** A ball is thrown upward from an initial height of 96 feet with an initial velocity of 16 feet per second. After t seconds, the height of the ball can be described by the expression $-16t^2 + 16t + 96$. What is the height of the ball after 3 seconds?

Writing & Thinking

13. Discuss like and unlike terms and give an example of each.

14. Explain the difference between -13^2 and $(-13)^2$.

Review Concepts

for *Voting and Apportionment*

10.R.1 Addition and Subtraction with Whole Numbers

10.R.2 Introduction to Decimal Numbers

10.R.1 Addition and Subtraction with Whole Numbers

Adding Whole Numbers

1. Write the numbers _____ so that the _____

2. Add only the _____

PROCEDURE

Carrying When Adding Whole Numbers

If the sum of the digits in one column is more than 9,

1. write the _____

2. carry the _____ as a number to be added to _____

 _____

PROCEDURE

Commutative Property of Addition

The order of the numbers in addition can be _____

For example, _____

DEFINITION

Associative Property of Addition

The grouping of the numbers in addition can be _____

For example, _____

DEFINITION

Additive Identity Property

The sum of a number and 0 is _____

For example, _____

The number 0 is called the _____

DEFINITION

1. The **perimeter** of a geometric figure is _____

▶ Watch and Work

Watch the video for Example 6 in the software and follow along in the space provided.

Example 6 Calculating the Perimeter of a Polygon

Calculate the perimeter of the polygon.

(**Note:** A 5-sided polygon is called a **pentagon**.)

Solution

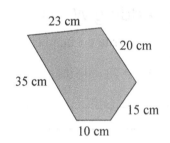

✏ Now You Try It!

Use the space provided to work out the solution to the next example.

Example A Calculating the Perimeter of a Polygon

Calculate the perimeter of the polygon.

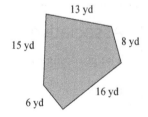

Subtraction

Subtraction is the operation of _____

The **difference** is the result of _____

_____

DEFINITION

Subtracting Whole Numbers

1. Write the numbers _____ so that the _____

2. Subtract only the _____

3. Check by _____ The sum must be _____

PROCEDURE

10.R.1 Exercises

Concept Check

True/False. Determine whether each statement is true or false. If a statement is false, explain how it can be changed so the statement will be true. (**Note:** There may be more than one acceptable change.)

1. A polygon is a geometric figure in a plane with two or more sides.

2. To find the perimeter of a rectangle, add the lengths of the four sides.

3. When subtracting, sometimes the digit being subtracted is larger than the digit it is being subtracted from and so "carrying" must occur.

4. If your bank account has a balance of $743 and you want to withdraw $115, you would use subtraction to find that the new balance would be $628.

Practice

Simplify.

5.   15
     +43

6.   981
     +46

7.   275
      −131

8.   543
      −167

9. Calculate the perimeter of the given geometric figure.

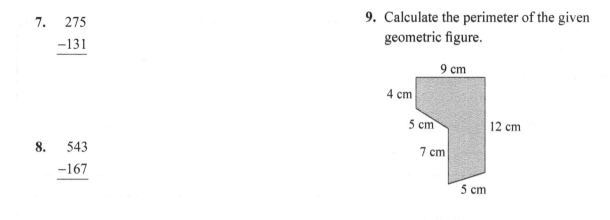

Applications

Solve.

10. *Budgeting:* The Magley family has the following monthly budget: $815 mortgage; $69 electric; $47 water; and $122 phone bills (including cell phones). What is the family's budget for each month for these expenses?

11. *Credit:* A couple sold their house for $135,000. They paid the realtor $8100, and other expenses of the sale came to $800. If they owed the bank $87,000 for the mortgage, what were their net proceeds from the sale?

Writing & Thinking

12. Explain when "carrying" should be used in addition with whole numbers and give an example.

13. Explain when "borrowing" would be used in subtraction and give an example.

10.R.2 Introduction to Decimal Numbers

To Read or Write a Decimal Number

1. Read (or write) the _____

2. Read (or write) the _____

3. Read (or write) the _____.

Then, name the fraction part with the _____.

PROCEDURE

To Compare Two Decimal Numbers

1. Moving left to right, compare digits _____

2. When one compared digit is larger, then the _____

PROCEDURE

▶ Watch and Work

Watch the video for Example 6 in the software and follow along in the space provided.

Example 6 Comparing Decimal Numbers

Arrange the following three decimal numbers in order from smallest to largest: 6.67, 5.14, 6.28. Then, graph them on a number line.

Solution

✏ Now You Try It!

Use the space provided to work out the solution to the next example.

Example A Comparing Decimal Numbers

Arrange the following three decimal numbers in order from smallest to largest: 2.6, 2.06, 2.46. Then, graph them on a number line.

Rounding Rule for Decimal Numbers

1. Look at the single digit one place value to the right of the digit in the place of desired accuracy.

   a. **If this digit is less than 5,** _____

   and replace all digits to the right with zeros. All digits to the _____.

   b. **If this digit is 5 or greater,** _____

   and replace all digits to the right with zeros. All digits to the _____

   _____. Then, the 9 is replaced by 0 and the _____

   _____.

2. Zeros to the right of the place of accuracy that are also to the right of the _____

   _____.

PROCEDURE

10.R.2 Exercises

Concept Check

True/False. Determine whether each statement is true or false. If a statement is false, explain how it can be changed so the statement will be true. (**Note:** There may be more than one acceptable change.)

   1. Two hundred thousand, four hundred six and twelve hundredths can be written as 200,406.12.

2. 92.586 is greater than 92.6.

3. On a number line, any number to the left of another number is larger than that other number.

4. When a decimal number is rounded, all numbers to the right of the place of accuracy become zeros in the final answer.

Practice

5. Write $2\dfrac{57}{100}$ in decimal notation.

6. Write 20.7 in words.

7. Write six and twenty-eight thousandths in decimal notation.

8. Arrange 0.2, 0.26, and 0.17 in order from smallest to largest. Then, graph the numbers on a number line.

Fill in the blanks to correctly complete each statement.

9. Round 3.00652 to the nearest ten-thousandth.

    a. The digit in the ten-thousandths position is ___.

    b. The next digit to the right is ___.

    c. Since ___ is less than 5, leave ___ as it is and replace ___ with 0.

    d. So 3.00652 rounds to _____ to the nearest ten-thousandth.

Applications

In each exercise, write the decimal numbers that are not whole numbers in words.

10. *Unicycles:* The tallest unicycle ever ridden was 114.8 feet tall, and was ridden by Sam Abrahams (with a safety wire suspended from an overhead crane) for a distance of 28 feet in Pontiac, Michigan, on January 29, 2004.[1]

11. *Water Weight:* One quart of water weighs approximately 2.0825 pounds.

Writing & Thinking

12. Discuss situations where you think it is particularly appropriate (or necessary) to write numbers in English word form.

13. With **a.** and **b.** as examples, explain in your own words how you can tell quickly when one decimal number is larger (or smaller) than another decimal number.

    a. The decimal number 2.765274 is larger than the decimal number 2.763895.

    b. The decimal number 17.345678 is larger than the decimal number 17.345578.

1 Source: http://semcycle.biz/record

CHAPTER 11.R

Review Concepts

for *The Arts*

11.R.1 Decimal Numbers and Fractions

To Change from Decimal Numbers to Fractions

A decimal number less than 1 (digits are to the right of the decimal point) can be written in fraction form by writing a fraction with:

1. A numerator that consists of the whole number formed by _____

   _____

2. A denominator that is the power of _____

   (For example, a denominator of 100 corresponds to _____.)

   PROCEDURE

1. _____ numbers can be **repeating** or **nonrepeating**. A **repeating decimal** has a repeating pattern to its digits.

2. Every fraction with a whole number numerator and nonzero denominator is _____

   _____. Such numbers are called _____.

▶ Watch and Work

Watch the video for Example 8 in the software and follow along in the space provided.

Example 8 Simplifying Expressions with Decimals and Fractions

Find the sum $10\dfrac{1}{2} + 7.32 + 5\dfrac{3}{5}$ in decimal form.

Solution

✏ Now You Try It!

Use the space provided to work out the solution to the next example.

Example A Simplifying Expressions with Decimals and Fractions

Find the sum $2.88 + \dfrac{1}{4} + 13\dfrac{9}{10}$ in decimal form.

11.R.1 Exercises

Concept Check

True/False. Determine whether each statement is true or false. If a statement is false, explain how it can be changed so the statement will be true. (**Note:** There may be more than one acceptable change.)

1. When a decimal number is changed to a fraction, the denominator will be the power of 10 that names the rightmost digit of the decimal number.

2. When a decimal number is changed to a fraction, the numerator can be determined by using the whole number that is formed by all the digits of the decimal number.

3. Fractions can always be converted to decimal form without losing accuracy.

4. In decimal form, $\dfrac{1}{3}$ is repeating and nonterminating.

Practice

Change each decimal number to a fraction or mixed number in lowest terms.

5. 0.18

6. 2.75

Change each fraction to a decimal number rounded to the nearest hundredth.

7. $\dfrac{20}{3}$

8. $\dfrac{40}{9}$

Simplify the expression by first writing all of the numbers in decimal form. Round to the nearest hundredth, if necessary.

9. $\dfrac{1}{4} + 0.25 + \dfrac{1}{5}$

10. Arrange $0.76, \dfrac{3}{4}, \dfrac{7}{10}$ in order from smallest to largest.

Applications

Solve.

11. *Rectangles:* A rectangle measures 6.4 inches in length, and has a width that measures $\frac{2}{5}$ of the length. Find the perimeter of the rectangle.

12. *Groceries:* A loaf of bread weighs 21.6 ounces. Mauricio cut off a third of the loaf to save for later and then cut the remaining portion into 16 equal slices. What was the weight of each slice of the 16 slices he cut?

Writing & Thinking

13. Describe the process used to change a terminating decimal number to a fraction.

14. List 2 different ways to solve this problem: $\frac{1}{2} + 3.67 - \frac{1}{8}$. State which method you prefer and why.

11.R.2 Ratios, Unit Rates, and Proportions

Ratios

A **ratio** is a comparison of _____

The ratio of a to b can be written as

_____

DEFINITION

Ratios have the following characteristics.

1. Ratios can be _____

2. The common units in a ratio can be _____

3. Generally, ratios are written with whole numbers in the _____

_____

A **rate** is a ratio with different _____

Changing Rates to Unit Rates

To make a rate a unit rate, _____

PROCEDURE

▶ Watch and Work

Watch the video for Example 9 in the software and follow along in the space provided.

Example 9 Application: Writing a Unit Rate

A bicyclist rides at a steady speed over level ground of 27 miles (mi) in 2.25 hours (hr). Find her speed in miles per hour.

Solution

✏️ Now You Try It!

Use the space provided to work out the solution to the next example.

Example A Application: Writing a Unit Rate

A swimmer swims laps at a steady pace of 6 laps in $\frac{1}{4}$ hour. Find his speed in laps per hour.

Proportions

A **proportion** is a statement that _____

In symbols,

_____

A proportion is true if _____

DEFINITION

To Solve a Proportion

1. Find the cross products (or cross multiply) and then _____

2. Divide both sides of the equation by _____

3. _____

PROCEDURE

▶ Watch and Work

Watch the video for Example 16 in the software and follow along in the space provided.

Example 16 Solving Proportions

Find the value of x if $\dfrac{4}{8} = \dfrac{5}{x}$.

Solution

✏ Now You Try It!

Use the space provided to work out the solution to the next example.

Example A Solving Proportions

Find the value of x if $\dfrac{12}{x} = \dfrac{9}{15}$.

To Solve an Application Using a Proportion

1. Identify the unknown quantity and _____

2. Set up a proportion in which the _____. (Make sure that
   the _____)

3. Solve the _____

PROCEDURE

11.R.2 Exercises

True/False. Determine whether each statement is true or false. If a statement is false, explain how it can be changed so the statement will be true. (**Note:** There may be more than one acceptable change.)

1. The units in the numerator and denominator of a ratio must be the same, or need to be able to be converted to the same units.

2. The order of the numbers in a ratio or a rate is irrelevant as long as the numbers are reduced.

3. The ratio 8:2 can be reduced to the ratio 4.

4. To make a unit rate, divide the numerator by the denominator.

5. A proportion is a statement that two ratios are being multiplied.

6. Cross canceling is used to determine if a proportion is true.

7. In order to solve the proportion $\dfrac{16}{36.8} = \dfrac{x}{27.6}$ we construct the equation $36.8x = 441.6$.

8. When using proportions to solve a word problem, there is only one correct way to set up the proportion.

9. The proportions $\dfrac{36 \text{ tickets}}{\$540} = \dfrac{x \text{ tickets}}{\$75}$ and $\dfrac{x \text{ tickets}}{36 \text{ tickets}} = \dfrac{\$75}{\$540}$ will yield the same answer.

Practice

10. Write 18 to 28 as a fraction in lowest terms.

11. Write 5 nickels to 3 quarters as a fraction in lowest terms by first finding common unites in the numerator and denominator.

12. Write the rate $200 in profit to $500 invested as a fraction in lowest terms.

13. Write the rate 270 miles to 4.5 hours as a unit rate.

Find the unit price (to the nearest tenth of a cent) of set of items and tell which one is the better (or best) purchase.

14. *Coffee beans:* 1.75 oz at $1.99, 12 oz at $7.99

Determine whether each proportion is true or false.

15. $\dfrac{3}{6} = \dfrac{4}{8}$

16. $\dfrac{1}{3} = \dfrac{33}{100}$

Solve each proportion.

17. $\dfrac{5}{4} = \dfrac{x}{8}$

18. $\dfrac{3.5}{2.6} = \dfrac{10.5}{B}$

Applications

Solve.

19. *Nutrition:* A serving of four home-baked chocolate chip cookies weighs 40 grams and contains 12 grams of fat. What is the ratio, in lowest terms, of fat grams to total grams?

20. **Standardized Testing:** In recent years, 18 out of every 100 students taking the SAT (Scholastic Aptitude Test) at a local school have scored 600 or above on the mathematics portion of the test. Write the ratio, in lowest terms, of the number of scores 600 or above to the number of scores below 600.

21. **Concrete:** The quality of concrete is based on the ratio of bags of cement to cubic yards of gravel. One batch of concrete consists of 27 bags of cement mixed into 9 cubic yards of gravel, while a second has 15 bags of cement mixed with 5 cubic yards of gravel. Determine whether the ratio of cement to gravel is the same for both batches.

22. **Grading:** An English teacher must read and grade 27 essays. If the teacher takes 20 minutes to read and grade 3 essays, how much time will he need to grade all 27 essays?

Writing & Thinking

23. Demonstrate three different ways the ratio comparing 5 apples to 3 apples can be written. Choose one form and explain why it is the preferred form when using ratios in a math course.

24. When finding price per unit, will monetary units be located in the numerator or the denominator of the rate?

25. In your own words, clarify how you can know that a proportion is set up correctly or not.

11.R.3 **Angles and Triangles**

Point, Line, Plane

Undefined Term	Representation	Discussion
Point		A point is represented by _____
		Points are labeled with _____
Line	_____	A line has no _____
		Lines are labeled with _____ _____
Plane	_____	Flat surfaces, such as a table top or wall, represent _____ _____
		Planes are labeled with _____
_____		

DEFINITION

Ray and Angle

Term	Definition	Illustrations with Notation
Ray	A ray consists of _____ _____ _____ _____	_____
Angle	An angle consists of _____ _____ _____ _____ _____	_____

DEFINITION

Labeling Angles

There are three common ways of labeling angles:

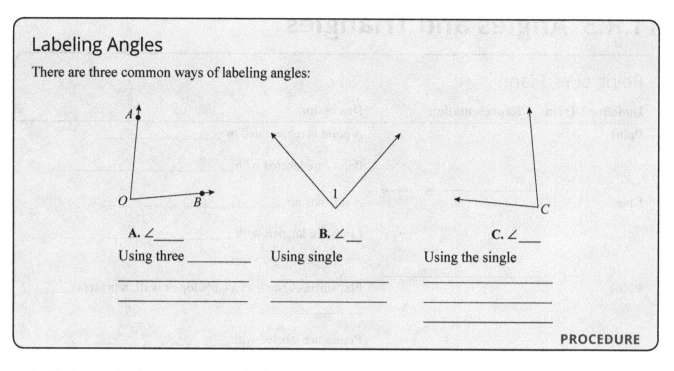

A. ∠____
Using three _____

B. ∠__
Using single

C. ∠___
Using the single

_____   _____   _____
_____   _____   _____
                                          _____

1. The base unit when measuring angles is _____ (symbolized _____).

Angles Classified by Measure

Name	Measure	Illustrations with Notation
Acute	_____	
		_____
Right	_____	
		_____
Obtuse	_____	
		_____
Straight	_____	
		_____

Complementary and Supplementary Angles

1. Two angles are **complementary** if _____

2. Two angles are **supplementary** if _____

DEFINITION

2. If two angles have the same measure, they are said to be _____ (symbolized

as ≅).

Vertical Angles

Vertical angles _____

That is, vertical angles have _____

DEFINITION

Adjacent Angles

Two angles are adjacent if _____

DEFINITION

Parallel Lines and Perpendicular Lines

Term	Definition	Illustrations with Notation
Parallel Lines	Two lines are parallel (symbolized ∥) if _____ _____ _____	$\overrightarrow{PQ}$ is parallel to ($\overrightarrow{PQ}$ _____)
Perpendicular Lines	Two lines are perpendicular (symbolized ⊥) if _____ _____ _____	$\overrightarrow{PQ}$ is perpendicular to ($\overrightarrow{PQ}$ _____)

DEFINITION

Parallel Lines and a Transversal

If two parallel lines are cut by a transversal, then the following two statements are true.

1. _____

2. _____

PROCEDURE

▶ Watch and Work

Watch the video for Example 8 in the software and follow along in the space provided.

Example 8 Calculating Measures of Angles

In the figure below, lines k and l are parallel, t is a transversal, and $m\angle 1 = 50°$. Find the measures of the other 7 angles.

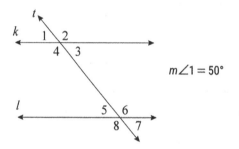

$$m\angle 1 = 50°$$

Solution

✏ Now You Try It!

Use the space provided to work out the solution to the next example.

Example A Calculating Measures of Angles

In the figure below, lines l and m are parallel, t is a transversal, and $m\angle 2 = 80°$. Find $m\angle 4$, $m\angle 5$, and $m\angle 6$.

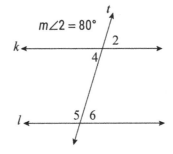

---

3. A **line segment** consists of _____

4. A **triangle** consists of _____.

Triangles Classified by Sides
(**Note:** In the figures, sides with equal length are indicated by the same number of tic marks.)

Name	Property	Example	
	_____		$\triangle ABC$ is scalene since _____
Scalene	_____		_____
	_____	_____	
	_____		$\triangle PQR$ is isosceles since _____
Isosceles	_____		
	_____	_____	
	_____		
Equilateral	_____		$\triangle XYZ$ is equilateral since _____
	_____	_____	

DEFINITION

Triangles Classified by Angles

Name	Property	Example
Acute	_____ _____ _____	$\triangle ABC$ is acute since _____ _____
Right	_____ _____ _____	$\triangle PRQ$ is a right triangle since _____
Obtuse	_____ _____ _____	$\triangle XYZ$ is an obtuse triangle since _____

DEFINITION

Three Properties of Triangles

In a triangle:

1. The sum of the measures _____

2. The sum of the lengths of _____

3. Longer sides are _____

PROPERTIES

11.R.3 **Exercises**

Concept Check

True/False. Determine whether each statement is true or false. If a statement is false, explain how it can be changed so the statement will be true. (**Note:** There may be more than one acceptable change.)

1. The sum of the measures of two complementary angles is equal to the measure of one right angle.

2. The sum of the measures of complementary angles is greater than the sum of the measures of supplementary angles.

3. Adjacent angles are two angles that share a side.

4. If two lines in a plane are not parallel, then they are perpendicular.

5. A triangle with sides of 4 inches, 4 inches, and 3 inches is an isosceles triangle.

6. A triangle with three angles that each measure less than 90 degrees is an acute triangle.

Practice

7. Name the type of angle formed by the hands on a clock.

a.

b.

c.

d.

   a. at six o'clock

   b. at three o'clock

   c. at one o'clock

   d. at five o'clock

8. Assume that $\angle 1$ and $\angle 2$ are complementary.

   a. If $m\angle 1 = 15°$, what is $m\angle 2$?

   b. If $m\angle 1 = 3°$, what is $m\angle 2$?

   c. If $m\angle 1 = 45°$, what is $m\angle 2$?

   d. If $m\angle 1 = 75°$, what is $m\angle 2$?

9. The figure shows two intersecting lines.

   a. If $m\angle 1 = 30°$, what is $m\angle 2$?

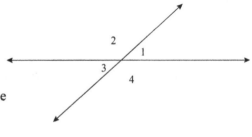

   b. Is $m\angle 3 = 30°$? Give a reason for your answer other than the fact that $\angle 1$ and $\angle 3$ are vertical angles.

   c. Name two pairs of congruent angles.

   d. Name four pairs of adjacent angles.

Classify each triangle in the most precise way possible, given the indicated lengths of its sides and/or measures of its angles.

10.

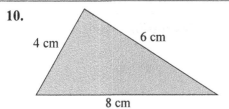

4 cm   6 cm

8 cm

11.

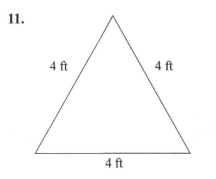

4 ft   4 ft

4 ft

Applications

Solve.

---

12. Suppose the lengths of the sides of △*DEF* are as shown in the figure. Is this possible? Explain your reasoning.

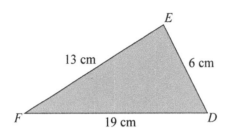

13. In the triangle shown, $m\angle X = 30°$ and $m\angle Y = 70°$.

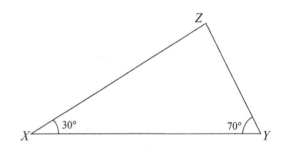

   a. What is $m\angle Z$?

   b. What kind of triangle is △*XYZ*?

   c. Which side is opposite $\angle X$?

   d. Which sides include $\angle X$?

   e. Is △*XYZ* a right triangle?

Writing & Thinking

14. Explain, in your own words, the relationships between vertex, ray, angle, and line.

11.R.4 **Rules for Exponents**

The Product Rule for Exponents

If a is a nonzero real number and m and n are integers, then

_____

In words, to multiply powers with the same base, _____

PROPERTIES

The Exponent 0

If a is a nonzero real number, then _____

The expression _____ is _____

DEFINITION

Quotient Rule for Exponents

If a is a nonzero real number and m and n are integers, then

_____

In words, to divide two powers with the same base, _____

PROPERTIES

Rule for Negative Exponents

If a is a nonzero real number and n is an integer, then

_____

PROPERTIES

▶ Watch and Work

Watch the video for Example 6 in the software and follow along in the space provided.

Example 6 Negative Exponents

Use the rule for negative exponents to simplify each expression so that it contains only positive exponents.

    a. 5^{-1}

    b. x^{-3}

    c. $x^{-9} \cdot x^{7}$

Solution

✏ Now You Try It!

Use the space provided to work out the solution to the next example.

Example A Negative Exponents

Use the rule for negative exponents to simplify each expression so that it contains only positive exponents.

    a. 7^{-1}

    b. x^{-7}

    c. $x^{-11} \cdot x^{6}$

Summary of the Rules for Exponents

For any nonzero real number a and integers m and n:

1. The exponent 1: _____

2. The exponent 0: _____

3. The product rule: _____

4. The quotient rule: _____

5. Negative exponents: _____

PROPERTIES

11.R.4 Exercises

Concept Check

True/False. Determine whether each statement is true or false. If a statement is false, explain how it can be changed so the statement will be true. (**Note:** There may be more than one acceptable change.)

1. If a constant does not have an exponent written, it is assumed that the exponent is 0.

2. If a is a nonzero real number and n is an integer, then $a^{-n} = -a^n$.

3. Since the product rule is stated for integer exponents, the rule is also valid for 0 and negative exponents.

4. When using the quotient rule, you should subtract the smaller exponent from the larger exponent.

Practice

Simplify each expression. The final form of the expressions with variables should contain only positive exponents. Assume that all variables represent nonzero numbers.

5. $y^3 \cdot y^8$

6. $\dfrac{y^7}{y^2}$

7. $x^{-3} \cdot x^0 \cdot x^2$

8. $\dfrac{10^4 \cdot 10^{-3}}{10^{-2}}$

Applications

Solve.

9. ***Computers:*** Rylee wants to move all her files to a new hard drive that has 2^{12} GB of storage on it. She wants to designate the same amount of storage for each of 2^4 projects. How much storage should be assigned to each project? Write your answer as a power of two.

10. ***Bacteria:*** Trey is studying patterns in bacteria. For a positive test result in his experiment, bacteria must grow in population at a minimum rate of 3^2 in 24 hours. If the initial population of the bacteria is 3^5 and his final measurement after 24 hours is 3^8, should he mark the test as positive or negative?

11.R.5 **Rationalizing Denominators**

> ## To Rationalize a Denominator Containing a Square Root or a Cube Root
>
> 1. If the denominator contains a square root, _____
>
> _____
>
> 2. If the denominator contains a cube root, _____
>
> _____
>
> **PROCEDURE**

▶ **Watch and Work**

Watch the video for Example 2 in the software and follow along in the space provided.

Example 2 Rationalizing Denominators Containing Cube Roots

Rationalize each denominator. Assume that all variables represent positive real numbers.

a. $\dfrac{3}{\sqrt[3]{5}}$

b. $\dfrac{7}{\sqrt[3]{32y}}$

Solution

✏ Now You Try It!

Use the space provided to work out the solution to the next example.

Example A Rationalizing Denominators Containing Cube Roots

Rationalize each denominator. Assume that all variables represent positive real numbers.

a. $\dfrac{4}{\sqrt[3]{7}}$

b. $\dfrac{5}{\sqrt[3]{9x^5}}$

> ## To Rationalize a Denominator Containing
> ## a Sum or Difference Involving Square Roots
>
> If the denominator of a fraction contains a sum or difference involving a square root, rationalize the denominator by multiplying both the numerator and denominator by the _____
> _____
>
> 1. If the denominator is of the form $a - b$, _____
>
> 2. If the denominator is of the form $a + b$, _____
>
> The new denominator will be the _____
> _____
>
> **PROCEDURE**

11.R.5 **Exercises**

Concept Check

True/False. Determine whether each statement is true or false. If a statement is false, explain how it can be changed so the statement will be true. (**Note:** There may be more than one acceptable change.)

1. The conjugate of $y - \sqrt{5}$ is $y + \sqrt{5}$.

2. To rationalize the denominator, multiply only the denominator by an expression that will result in a denominator with no radicals.

3. To rationalize a fraction whose denominator is $\sqrt[3]{a}$, you would need to multiply the numerator and the denominator by $\sqrt[3]{a}$.

4. The fraction $\dfrac{\sqrt{2}}{3}$ is in simplest form.

Practice

Rationalize the denominator and simplify, if possible. Assume that all variables represent positive real numbers.

5. $\dfrac{-3}{\sqrt{7}}$

7. $\dfrac{2}{\sqrt{6} - 2}$

6. $\dfrac{\sqrt{ab}}{\sqrt{9ab}}$

8. $\dfrac{x}{\sqrt{x} + 2}$

Applications

Solve.

9. *Volume of a Cylinder:* The radius of a cylinder can be expressed in terms of its volume and its height by $r = \sqrt{\dfrac{V}{\pi h}}$. Rationalize the denominator of this formula.

10. *Business:* A company that sells computers learns that their income can be represented by the equation $I = \dfrac{6500p}{\sqrt{p}-10}$ dollars when they sell their computers for p dollars. Rationalize the denominator of this equation.

Writing & Thinking

11. In your own words, explain how to rationalize the denominator of a fraction containing the sum or difference of square roots in the denominator. Why does this work?

11.R.6 Quadratic Equations: The Quadratic Formula

> ## The Quadratic Formula
>
> For the general quadratic equation _____
>
> _____
>
> **FORMULA**

▶ Watch and Work

Watch the video for Example 2 in the software and follow along in the space provided.

Example 2 The Quadratic Formula

Solve by using the quadratic formula.

$7x^2 - 2x + 1 = 0$

Solution

✏ Now You Try It!

Use the space provided to work out the solution to the next example.

Example A The Quadratic Formula

Solve by using the quadratic formula.

$5x^2 - 3x + 2 = 0$

---

1. The expression $b^2 - 4ac$, the part of the quadratic formula that lies under the radical sign, is called

   the _____.

Discriminant	Nature of Solutions
$b^2 - 4ac > 0$	_____
$b^2 - 4ac = 0$	_____
$b^2 - 4ac < 0$	_____

Table 1

11.R.6 Exercises

Concept Check

True/False. Determine whether each statement is true or false. If a statement is false, explain how it can be changed so the statement will be true. (**Note:** There may be more than one acceptable change.)

1. The quadratic formula will always work when solving quadratic equations.

2. If the discriminant is a perfect square, the quadratic equation is factorable.

3. When using the quadratic formula, if the discriminant is greater than zero, there are infinite solutions.

4. If the discriminant is less than zero, there is no real solution.

Practice

Find the discriminant and determine the nature of the solutions of each quadratic equation.

5. $x^2 + 6x - 8 = 0$

6. $x^2 - 8x + 16 = 0$

Solve each of the quadratic equations using the quadratic formula.

7. $x^2 + 4x - 4 = 0$

8. $x^2 - 2x + 7 = 0$

9. $3x^2 - 7x + 4 = 0$

Applications

Solve.

10. ***Throwing Objects:*** 🧮 An orange is thrown down from the top of a building that is 300 feet tall with an initial velocity of 6 feet per second. The distance of the object from the ground can be calculated using the equation $d = 300 - 6t - 16t^2$, where t is the time in seconds after the orange is thrown.

   a. On a balcony, a cup is sitting on a table located 100 feet from the ground. If the orange is thrown with the right aim to fall into the cup, how long will the orange fall? Round to the nearest hundredth. (**Hint:** The distance is 100 feet.)

   b. If the orange misses the cup and falls to the ground, how long will it take for the orange to splatter on the sidewalk? (**Hint:** What is the height of the orange when it hits the ground?)

   c. Approximately how much longer would it take for the orange to fall to the sidewalk than it would for the orange to fall into the cup?

Writing & Thinking

11. Find an equation of the form $Ax^4 + Bx^2 + C = 0$ that has the four roots ± 2 and ± 3. Explain how you arrived at this equation.

12. The surface area of a right circular cylinder can be found using the following formula: $S = 2\pi r^2 + 2\pi rh$, where r is the radius of the cylinder and h is the height. Estimate the radius of a circular cylinder of height 30 cm and surface area 300 cm². Explain how you used your knowledge of quadratic equations.

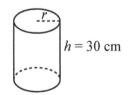
$h = 30$ cm

Review Concepts

for *Sports*

12.R.1 **Exponents and Order of Operations**

1. When looking at $3^5 = 243$, 3 is the _____, 5 is the _____, and 243

   is the _____. _____ are written slightly to the right and above

   the _____. The expression 3^5 is _____.

The Exponent 1

Any number raised to the first power _____

For example, _____

DEFINITION

The Exponent 0

Any nonzero number raised to the 0 power _____

For example, _____

Note: The expression 0^0 _____

DEFINITION

Rules for Order of Operations

1. Simplify within grouping symbols, such as _____
   (If there are more than one pair of grouping symbols, start with _____

   _____

2. Evaluate any _____

3. Moving from left to right, perform any _____

   _____

4. Moving from left to right, perform any _____

   _____

PROCEDURE

---

2. A well-known mnemonic device for remembering the rules for order of operations is the following.

Please	Excuse	My	Dear	Aunt	Sally
↓	↓	↓	↓	↓	↓
_____	_____	_____	_____	_____	_____

▶ Watch and Work

Watch the video for Example 6 in the software and follow along in the space provided.

Example 6 Using the Order of Operations with Whole Numbers

Simplify: $2 \cdot 3^2 + 18 \div 3^2$

Solution

✏ Now You Try It!

Use the space provided to work out the solution to the next example.

Example A Using the Order of Operations with Whole Numbers

Simplify: $6^2 \div 9 + 3 - 14 \div 7$

12.R.1 **Exercises**

Concept Check

True/False. Determine whether each statement is true or false. If a statement is false, explain how it can be changed so the statement will be true. (**Note:** There may be more than one acceptable change.)

1. Nine squared is equal to eighteen.

2. $2^7 = 128$

3. 7^0 is undefined.

4. According to the order of operations, multiplication is always performed before division.

Practice

For each exponential expression **a.** identify the base, **b.** identify the exponent, and **c.** evaluate the exponential expression.

5. 2^3

6. 4^0

Simplify.

7. $18 \div 2 - 1 - 3 \cdot 2$

8. $30 \div 2 - 11 + 2(5-1)^3$

Applications

Solve.

9. *Card Games:* Neville bought 15 boxes of trading cards. Each box has 10 packs of trading cards. Each pack of trading cards contains 20 cards. He adds 132 cards that he already owns to the newly purchased cards. Then, Neville evenly distributes all of the cards to 6 of his friends. How many trading cards would each person get?

   a. If you simplify the expression $15 \cdot 10 \cdot 20 + 132 \div 6$ using the order of operations, will you get the correct answer? If not, explain what is wrong with the expression.

   b. What is the answer? If necessary, write the corrected expression to get the correct results when following the order of operations.

10. *Purchases:* Robert is purchasing shirts for his weekend soccer team. The shirts he wants to buy are normally $25 each but are on sale for $10 off. His team has a total of 11 players. How much will he spend to buy the shirts?

    a. If you simplify the expression $\$25 - \$10 \cdot 11$ using the order of operations, will you get the correct answer? If not, explain what is wrong with the expression.

    b. What is the answer? If necessary, write the corrected expression to get the correct results when following the order of operations.

Writing & Thinking

11. Give one example where addition should be completed before multiplication.

12.R.2 **Ratios, Unit Rates, and Proportions**

> ## Ratios
>
> A **ratio** is a comparison of _____
> The ratio of a to b can be written as
>
> _____
>
> **DEFINITION**

Ratios have the following characteristics.

1. Ratios can be _____

2. The common units in a ratio can be _____

3. Generally, ratios are written with whole numbers in the _____

    _____

A **rate** is a ratio with different _____

> ## Changing Rates to Unit Rates
>
> To make a rate a unit rate, _____
>
> **PROCEDURE**

▶ **Watch and Work**

Watch the video for Example 9 in the software and follow along in the space provided.

Example 9 Application: Writing a Unit Rate

A bicyclist rides at a steady speed over level ground of 27 miles (mi) in 2.25 hours (hr). Find her speed in miles per hour.

Solution

✏ Now You Try It!

Use the space provided to work out the solution to the next example.

Example A Application: Writing a Unit Rate

A swimmer swims laps at a steady pace of 6 laps in $\frac{1}{4}$ hour. Find his speed in laps per hour.

Proportions

A **proportion** is a statement that _____

In symbols,

_____

A proportion is true if _____

DEFINITION

To Solve a Proportion

1. Find the cross products (or cross multiply) and then _____

2. Divide both sides of the equation by _____

3. _____

PROCEDURE

▶ **Watch and Work**

Watch the video for Example 16 in the software and follow along in the space provided.

___

Example 16   Solving Proportions

Find the value of x if $\dfrac{4}{8} = \dfrac{5}{x}$.

Solution

___

✏ **Now You Try It!**

Use the space provided to work out the solution to the next example.

___

Example A Solving Proportions

Find the value of x if $\dfrac{12}{x} = \dfrac{9}{15}$.

___

To Solve an Application Using a Proportion

1. Identify the unknown quantity and _____

2. Set up a proportion in which the _____. (Make sure that
    the _____)

3. Solve the _____

PROCEDURE

12.R.2 Exercises

True/False. Determine whether each statement is true or false. If a statement is false, explain how it can be changed so the statement will be true. (**Note:** There may be more than one acceptable change.)

1. The units in the numerator and denominator of a ratio must be the same, or need to be able to be converted to the same units.

2. The order of the numbers in a ratio or a rate is irrelevant as long as the numbers are reduced.

3. The ratio 8:2 can be reduced to the ratio 4.

4. To make a unit rate, divide the numerator by the denominator.

5. A proportion is a statement that two ratios are being multiplied.

6. Cross canceling is used to determine if a proportion is true.

7. In order to solve the proportion $\dfrac{16}{36.8} = \dfrac{x}{27.6}$ we construct the equation $36.8x = 441.6$.

8. When using proportions to solve a word problem, there is only one correct way to set up the proportion.

9. The proportions $\dfrac{36 \text{ tickets}}{\$540} = \dfrac{x \text{ tickets}}{\$75}$ and $\dfrac{x \text{ tickets}}{36 \text{ tickets}} = \dfrac{\$75}{\$540}$ will yield the same answer.

Practice

10. Write 18 to 28 as a fraction in lowest terms.

12. Write the rate $200 in profit to $500 invested as a fraction in lowest terms.

11. Write 5 nickels to 3 quarters as a fraction in lowest terms by first finding common unites in the numerator and denominator.

13. Write the rate 270 miles to 4.5 hours as a unit rate.

Find the unit price (to the nearest tenth of a cent) of set of items and tell which one is the better (or best) purchase.

14. *Coffee beans:* 1.75 oz at $1.99, 12 oz at $7.99

Determine whether each proportion is true or false.

15. $\dfrac{3}{6} = \dfrac{4}{8}$

16. $\dfrac{1}{3} = \dfrac{33}{100}$

Solve each proportion.

17. $\dfrac{5}{4} = \dfrac{x}{8}$

18. $\dfrac{3.5}{2.6} = \dfrac{10.5}{B}$

Applications

Solve.

19. *Nutrition:* A serving of four home-baked chocolate chip cookies weighs 40 grams and contains 12 grams of fat. What is the ratio, in lowest terms, of fat grams to total grams?

20. *Standardized Testing:* In recent years, 18 out of every 100 students taking the SAT (Scholastic Aptitude Test) at a local school have scored 600 or above on the mathematics portion of the test. Write the ratio, in lowest terms, of the number of scores 600 or above to the number of scores below 600.

21. *Concrete:* The quality of concrete is based on the ratio of bags of cement to cubic yards of gravel. One batch of concrete consists of 27 bags of cement mixed into 9 cubic yards of gravel, while a second has 15 bags of cement mixed with 5 cubic yards of gravel. Determine whether the ratio of cement to gravel is the same for both batches.

22. *Grading:* An English teacher must read and grade 27 essays. If the teacher takes 20 minutes to read and grade 3 essays, how much time will he need to grade all 27 essays?

Writing & Thinking

23. Demonstrate three different ways the ratio comparing 5 apples to 3 apples can be written. Choose one form and explain why it is the preferred form when using ratios in a math course.

24. When finding price per unit, will monetary units be located in the numerator or the denominator of the rate?

25. In your own words, clarify how you can know that a proportion is set up correctly or not.

12.R.3 Simplifying and Evaluating Algebraic Expressions

Like Terms

Like terms (or similar terms) are terms that are _____

_____

DEFINITION

Combining Like Terms

To combine like terms, _____

DEFINITION

To Evaluate an Algebraic Expression

1. _____

2. _____

3. _____

(**Note:** Terms separated by + and – signs may be evaluated _____

_____.)

PROCEDURE

▶ Watch and Work

Watch the video for Example 5 in the software and follow along in the space provided.

Example 5 Simplifying and Evaluating Algebraic Expressions

Simplify and evaluate $3ab - 4ab + 6a - a$ for $a = 2$, $b = -1$.

Solution

✎ Now You Try It!

Use the space provided to work out the solution to the next example.

Example A Identifying Properties of Addition and Multiplication

Simplify and evaluate

$5ab - 8ab + 2a - 3a$ for $a = -3$, $b = 1$

12.R.3 Exercises

True/False. Determine whether each statement is true or false. If a statement is false, explain how it can be changed so the statement will be true. (**Note:** There may be more than one acceptable change.)

1. A variable that does not appear to have an exponent has an exponent of 1.

2. In the term $-9x$, nine is being subtracted from x.

3. In the term "$12a$," 12 is the constant.

4. Like terms have the same coefficients.

Practice

Identify the like terms in each list of terms.

5. $-5, 3, 7x, 8, 9x, 3y$

Simplify each expression by combining like terms.

6. $8x + 7x$

7. $3x - 5x + 12x$

8. $13x + 12x^2 + 15x - 35 - 41 - 2x^2$

Simplify each expression and then evaluate the expression for $y = 3$ and $a = -2$.

9. $5y + 4 - 2y$

10. $\dfrac{3a + 5a}{-2} + 12a$

Applications

Solve.

11. ***Profit:*** An apartment management company owns a property with 100 units. The company has determined that the profit made per month from the property can be calculated using the equation $P = -10x^2 + 1500x - 6000$, where x is the number of units rented per month. How much profit does the company make when 80 units are rented?

12. ***Physics:*** A ball is thrown upward from an initial height of 96 feet with an initial velocity of 16 feet per second. After t seconds, the height of the ball can be described by the expression $-16t^2 + 16t + 96$. What is the height of the ball after 3 seconds?

Writing & Thinking

13. Discuss like and unlike terms and give an example of each.

14. Explain the difference between -13^2 and $(-13)^2$.

12.R.4 US Measurements

Using Multiplication and Division to Convert Measurements

1. Multiply to convert to _____ (There will be _____.)

2. Divide to convert to _____. (There will be _____.)

PROCEDURE

Using Unit Fractions to Convert Measurements

1. The numerator should be in the _____

2. The denominator should be in the _____

PROCEDURE

▶ Watch and Work

Watch the video for Example 6 in the software and follow along in the space provided.

Example 6 Application: Converting US Units of Measure

Determine how many seconds are in a 5-day work week assuming an 8 hr work day.

Solution

✏ Now You Try It!

Use the space provided to work out the solution to the next example.

Example A Application: Finding the Mean

How many fluid ounces are in 8 gallons of apple juice?

12.R.4 Exercises

Concept Check

True/False. Determine whether each statement is true or false. If a statement is false, explain how it can be changed so the statement will be true. (**Note:** There may be more than one acceptable change.)

1. Capacity can be measured using ounces, quarts, and gallons.

2. One mile is equivalent to 2000 feet.

3. To convert from smaller units to larger units, division will be required.

4. Multiplication by a unit fraction does not change the value of the expressions being converted.

Practice

Convert each measurement.

5. 4 pt = ____ c

6. 10 mi = ____ ft

7. 39 ft = ____ yd

8. 150 min = ____ hr

Applications

Solve.

9. *Interior Decorating:* Sheer fabric costs $7.99 per yard. If it will take 35 feet of fabric to make drapes for the entire house, how much must you spend on fabric for the drapes, to the nearest cent?

10. *Publishing:* The author of this textbook spent 1 year, 23 weeks, 5 days, and 14 hours writing it. How many seconds is this? (**Hint:** There are 52 weeks in a year.)

Writing & Thinking

11. Colby needs to find out how many yards are in one mile. What two sets of equivalent units would he need to make that determination?

12. In your own words, explain when you would multiply and when you would divide when converting between units.

12.R.5 The Metric System: Length and Area

Writing Metric Units of Measure

In the metric system,

1. A 0 is written to the left of the decimal point if _____

   For example, _____

2. No commas are used in writing numbers. If a number has more than four digits (to the left or right of the decimal point), the digits are _____

   _____

   For example, _____

   PROCEDURE

1. There are two basic methods of converting units of measurement in the metric system:

   1. multiplying by_____,

   2. moving the _____

Using Unit Fractions to Convert Measures

1. The numerator should be in the _____

2. The denominator should be in the _____

   PROCEDURE

▶ Watch and Work

Watch the video for Example 6 in the software and follow along in the space provided.

Example 6 Converting Metric Units of Area

Convert each measurement using unit fractions.

a. $5 \text{ cm}^2 = $ _____ mm^2

b. $4600 \text{ mm}^2 = $ _____ m^2

Solution

✏ Now You Try It!

Use the space provided to work out the solution to the next example.

Example A Converting Metric Units of Area

Convert each measurement using unit fractions.

a. $86 \text{ m}^2 = $ _____ cm^2

b. $0.06 \text{ mm}^2 = $ _____ dm^2

12.R.5 Exercises

Concept Check

True/False. Determine whether each statement is true or false. If a statement is false, explain how it can be changed so the statement will be true. (**Note:** There may be more than one acceptable change.)

1. To change from smaller units to larger units, multiplication must be used.

2. Units of length in the metric system are named by putting a prefix in front of the basic unit meter, for example, centimeter.

3. In metric units, a square that is 1 centimeter long on each side is said to have an area of 1 centimeter.

Practice

Convert each measurement.

4. 3 m = ___ cm

5. 19.77 m = ___km

6. 6 500 000 hertz = ___ megahertz

7. 13 dm² = _____ cm² = _____ mm²

Applications

Solve.

8. *Geometry:* A triangle has a base measuring 4 cm and a height measuring 16 mm. Determine the area of the triangle in cm².

9. *Transportation:* A section of railroad track measuring 2.1 km in length needs to be replaced. Each railroad tie is 4 decimeters wide and they are to be spaced 0.8 m apart. How many railroad ties will be needed to complete this section of track?

Writing & Thinking

10. Compare and contrast ease of converting units in the US customary system and the metric system.

11. Discuss the meaning of prefixes like milli-, centi-, and kilo- in metric units. Give examples.

12.R.6 US and Metric Equivalents

Temperature

US customary measure is in _____

Metric measure is in _____

DEFINITION

Temperature Formulas

F = Fahrenheit temperature   and C = Celsius temperature

$F = $ _____ $C = $ _____

FORMULA

▶ Watch and Work

Watch the video for Example 5 in the software and follow along in the space provided.

Example 5   Converting Units of Area

Convert each measurement, rounding to the nearest hundredth.

a. $40 \text{ yd}^2 = $ _____ m^2

b. $100 \text{ cm}^2 = $ _____ in.^2

c.   6 acres = _____ ha

d.   5 ha = _____ acres

Solution

✎ Now You Try It!

Use the space provided to work out the solution to the next example.

Example A Converting Units of Area

Convert each measurement, rounding to the nearest hundredth.

a. $53 \text{ in.}^2 = \underline{\hspace{2cm}} \text{ cm}^2$

b. $50 \text{ m}^2 = \underline{\hspace{2cm}} \text{ ft}^2$

c. $16 \text{ acres} = \underline{\hspace{2cm}} \text{ ha}$

d. $3 \text{ ha} = \underline{\hspace{2cm}} \text{ acres}$

12.R.6 Exercises

Concept Check

True/False. Determine whether each statement is true or false. If a statement is false, explain how it can be changed so the statement will be true. (**Note:** There may be more than one acceptable change.)

1. Water freezes at 32 degrees Celsius.

2. When converting between US customary and metric units, often the results will be approximations.

3. A 5K (km) run is longer than a 5 mile run.

4. One square meter covers more area than one square yard.

Practice

Convert each measurement. Round to the nearest hundredth if necessary.

5. 25 °C = ___ °F

6. 9 ft = ___ m

7. 3 in.² = _____ cm²

8. 4 qt = ___ L

9. 33 kg = ___ lb

Applications

Solve.

10. ***Baking:*** While visiting her aunt in Germany, Helga wants to surprise her aunt with a cake. She brought her mom's cake recipe with her from Georgia. The recipe says to bake the cake at 350 degrees Fahrenheit but the temperature gauge on her aunt's oven is in degrees Celsius. To what temperature should Helga set her aunt's oven in order to bake the cake at the correct temperature? Round the temperature to the nearest degree.

11. ***Sports:*** The Ironman Triathlon championship in Hawaii consists of a swim of 3.86 km, a bike ride of 180.25 km, and finishes with a run equal to the length of a standard marathon. A marathon is typically 26.2 miles. What is the total length of the Ironman Triathlon in kilometers? Round the length to the nearest tenth of a km.

Writing & Thinking

12. Most conversions between the US customary system of measure and metric system are not exact. Explain why this is true and give any exceptions.

Review Concepts

for *Graph Theory*

13.R.1 Solving Linear Equations: $ax + b = c$

Procedure for Solving Linear Equations that
Simplify to the Form $ax + b = c$

1. Combine _____

2. Use the **addition principle of equality** and _____

   _____

3. Use the _____ and multiply both sides of the

   equation by the reciprocal of the coefficient of the variable (**or** _____

   _____ **itself**). The _____ will become +1.

4. Check your answer by _____

   PROCEDURE

▶ Watch and Work

Watch the video for Example 2 in the software and follow along in the space provided.

Example 2 Solving Linear Equations of the Form $ax + b = c$

Solve the equation: $-26 = 2y - 14 - 4y$

Solution

✎ Now You Try It!

Use the space provided to work out the solution to the next example.

Example A Solving Linear Equations of the Form *ax + b = c*

Solve the equation.

$-18 = 2y - 8 - 7y$

13.R.1 Exercises

Concept Check

True/False. Determine whether each statement is true or false. If a statement is false, explain how it can be changed so the statement will be true. (**Note:** There may be more than one acceptable change.)

1. If an equation of the form $ax + b = c$ uses decimal or fractional coefficients, the addition and multiplication principles of equality cannot be used.

2. The first step in solving $2x + 3 = 9$ is to add 3 to both sides.

3. To solve an equation that has been simplified to $4x = 12$, you need to multiply both sides by $\frac{1}{4}$, or divide both sides by 4.

4. When solving a linear equation with decimal coefficients, one approach is to multiply both sides in such a way to give integer coefficients before solving.

Practice

Solve each equation.

5. $3x + 11 = 2$

6. $-5x + 2.9 = 3.5$

7. $\dfrac{2}{5} - \dfrac{1}{2}x = \dfrac{7}{4}$

8. $\dfrac{y}{3} - \dfrac{2}{3} = 7$

Applications

Solve.

9. *Music:* The tickets for a concert featuring the new hit band, Flying Sailor, sold out in 2.5 hours. If there were 35,000 tickets sold, solve the equation $35,000 - 2.5x = 0$ to find the number of tickets sold per hour.

10. *Movies:* All snacks (candy, popcorn, and soda) cost $3.50 each at the local movie theater. Admission tickets cost $7.50 each. After a long week, Carlos treats himself to a night at the movies. His movie night budget is $25 and he spends all his movie money. Solve the equation $3.50x + $7.50 = $25.00 to determine how many snacks Carlos can buy.

Writing & Thinking

11. Find the error(s) made in solving each equation and give the correct solution.

a. $\dfrac{1}{3}x + 4 = 9$

$3 \cdot \dfrac{1}{3}x + 4 = 3 \cdot 9$

$x + 4 = 27$

$x + 4 - 4 = 27 - 4$

$x = 23$

b. $5x + 3 = 11$

$(5x - 3) + (3 - 3) = 11 - 3$

$2x + 0 = 8$

$\dfrac{2x}{2} = \dfrac{8}{2}$

$x = 4$

13.R.2 The Real Number Line and Absolute Value

Integers

The set of numbers consisting of the _____

_____

DEFINITION

Variables

A **variable** is a symbol (generally a _____) that is used to _____

_____

DEFINITION

Rational Numbers

A **rational number** is a number that can be written in

_____

_____

OR

A **rational number** is a number that can be written in _____

_____

DEFINITION

---

1. Numbers that cannot be written as fractions with integer numerators and denominators are

   called _____

▶ Watch and Work

Watch the video for Example 4 in the software and follow along in the space provided.

Example 4 Graphing Sets of Numbers

Graph the set of **real numbers** $\left\{-\dfrac{3}{4}, 0, 1, 1.5, 3\right\}$.

Solution

✏ Now You Try It!

Use the space provided to work out the solution to the next example.

Example A   Graphing Sets of Numbers

Graph the set of real numbers

$$\left\{-2.5, -1, 0, \frac{5}{4}, 4\right\}.$$

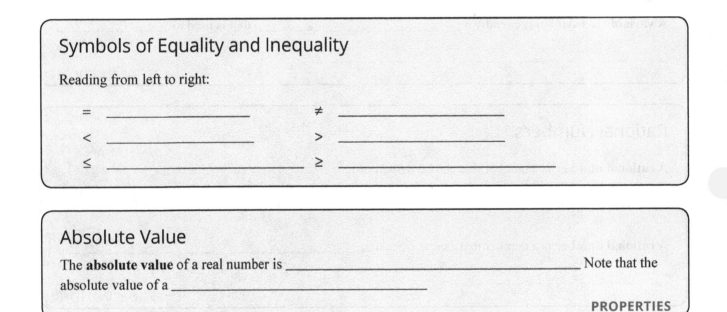

Symbols of Equality and Inequality

Reading from left to right:

= _____   ≠ _____

< _____   > _____

≤ _____   ≥ _____

Absolute Value

The **absolute value** of a real number is _____ Note that the

absolute value of a _____

PROPERTIES

13.R.2 Exercises

Concept Check

True/False. Determine whether each statement is true or false. If a statement is false, explain how it can be changed so the statement will be true. (**Note:** There may be more than one acceptable change.)

1. On a number line, smaller numbers are always to the left of larger numbers.

2. The absolute value of a negative number is a positive number.

3. All whole numbers are also integers.

4. Zero is a positive number.

Graph each set of real numbers on a real number line.

5. $\{-3, -2, 0, 1\}$

6. $\left\{-2, -1, -\dfrac{1}{3}, 2\right\}$

List the numbers in the set $A = \left\{-7, -\sqrt{6}, -2, -\dfrac{5}{3}, -1.4, 0, \dfrac{3}{5}, \sqrt{5}, \sqrt{11}, 4, 5.9, 8\right\}$ that are described in each exercise.

7. Whole numbers

8. Rational numbers

Determine whether each statement is true or false. If a statement is false, rewrite it in a form that is a true statement. (There may be more than one way to correct a statement.)

9. $0 = -0$

10. $|-8| \geq 4$

Applications

Solve. Represent each quantity with a signed integer.

11. *Oceans:* The Alvin is a manned deep-ocean research submersible that has explored the wreck of the Titanic. The operating depth of the Alvin is 4500 meters below sea level.

12. *Oceans:* The Mariana trench is the deepest known location on the Earth's ocean floor. The deepest known part of the Mariana Trench is approximately 11 kilometers below sea level.

Writing & Thinking

13. Explain, in your own words, how an expression such as $-y$ might represent a positive number.

14. Compare and contrast absolute value with opposites.

Review Concepts

for *Number Theory*

14.R.1 Multiplication with Whole Numbers

Commutative Property of Multiplication

The order of the numbers in multiplication can be _____

For example, _____

PROPERTIES

Associative Property of Multiplication

The grouping of the numbers in multiplication can be _____

For example, _____

PROPERTIES

Multiplicative Identity Property

The product of any number and 1 is _____

For example, _____

The number 1 is called the _____

PROPERTIES

Multiplication Property of 0 (or Zero-Factor Law)

The product of a number and 0 is _____

For example, _____

PROPERTIES

The Distributive Property

Multiplication can be _____

For example, _____

PROPERTIES

▶ **Watch and Work**

Watch the video for Example 5 in the software and follow along in the space provided.

Example 5 Multiplying Whole Numbers

Multiply: $93 \cdot 46$

Solution

The standard form of multiplication is used here to find the product $93 \cdot 46$.

✐ **Now You Try It!**

Use the space provided to work out the solution to the next example.

Example A Multiplying Whole Numbers

Multiply: 15
          $\times 32$

Multiplying Whole Numbers by Powers of 10

To multiply a whole number:

by 10, write _____

by 100, write _____

by 1000, write _____

by 10,000, write _____

and so on.

PROCEDURE

Area of a Rectangle

The **area** of a rectangle (measured in square units) is found by _____

DEFINITION

14.R.1 **Exercises**

Concept Check

True/False. Determine whether each statement is true or false. If a statement is false, explain how it can be changed so the statement will be true. (**Note:** There may be more than one acceptable change.)

1. The numbers being multiplied are called the divisors.

2. According to the multiplicative identity, $1 \cdot 25 = 52$.

3. According to the distributive property, $4 \cdot (7 + 2) = 4 \cdot 7 + 4 \cdot 2$.

4. The associative property of multiplication indicates that length can be multiplied by width or width can be multiplied by length to get the same answer.

Practice

5. Multiply: $\begin{array}{r} 42 \\ \times 56 \\ \hline \end{array}$

6. $40 \cdot 2000$

State the property of multiplication illustrated and show that the statement is true by performing the multiplication.

7. $3 \cdot (1 \cdot 7) = (3 \cdot 1) \cdot 7$

8. Rewrite $7(8 + 4)$ by using the distributive property then simplify.

9. Calculate the area of the given rectangle.

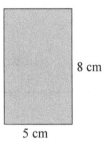

8 cm

5 cm

Applications

Solve.

10. *Television:* A network television station has approximately 18 minutes of commercial time in each hour. How many minutes of commercial time does the network have in a one-day programming schedule of 20 hours? In one week?

11. *Inventory:* A sandwich shop buys 372 loaves of bread for the week. If each loaf of bread has 24 slices, how many slices of bread were purchased?

Writing & Thinking

12. Explain, in your own words, what the zero-factor law indicates.

13. Explain, in your own words, why 1 is called the multiplicative identity.

14.R.2 Division with Whole Numbers

1. We can use the division sign (÷) to indicate the division procedure as follows.

$$12 \quad ÷ \quad 4 \quad = \quad 3 \qquad \text{Read "12 divided by 4 equals 3."}$$

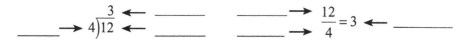

_____ ÷ _____ = _____

2. Two other notations that indicate division are the following.

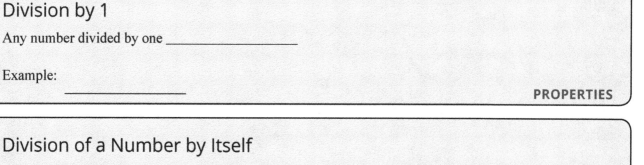

$$\_\_\_\_ \rightarrow 4\overline{)12} \leftarrow \_\_\_\_ \qquad \_\_\_\_ \rightarrow \frac{12}{4} = 3 \leftarrow \_\_\_\_$$

Division by 1

Any number divided by one _____

Example:

    _____

PROPERTIES

Division of a Number by Itself

Any nonzero number _____

Example:

    _____

PROPERTIES

Division Involving 0

Case 1: Any nonzero whole number _____

Example:

    _____

Case 2: _____

Example:

    _____

PROPERTIES

---

3. The long division process can be written in the following format.

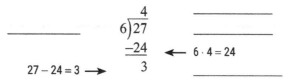

$$_____ \qquad 6\overline{)27}^{\,4} \qquad _____$$

$$27 - 24 = 3 \rightarrow \quad \frac{-24}{3} \quad \leftarrow 6 \cdot 4 = 24$$

▶ Watch and Work

Watch the video for Example 3 in the software and follow along in the space provided.

Example 3 Dividing Whole Numbers

Divide: 683 ÷ 7

Solution

✏ Now You Try It!

Use the space provided to work out the solution to the next example.

Example A Dividing Whole Numbers

Divide: 415 ÷ 6

14.R.2 **Exercises**

Concept Check

True/False. Determine whether each statement is true or false. If a statement is false, explain how it can be changed so the statement will be true. (**Note:** There may be more than one acceptable change.)

  1. If a division problem has a nonzero remainder, then the divisor and quotient are factors of the dividend.

  2. $13 \div 1 = 13$

  3. $12 \div 0 = 12$

  4. $\dfrac{0}{7}$ is undefined.

Practice

Divide.

  5. $13\overline{)0}$

  6. $0\overline{)51}$

  7. $12\overline{)108}$

  8. $11\overline{)4406}$

Applications

Solve.

9. *Nutrition Facts:* One pint of Ben and Jerry's Crème Brûlée Ice Cream has 64 grams of fat. If there are 4 servings per pint, how many grams of fat are in each serving?

10. *Space Travel:* US Astronaut Peggy Whitson orbited the Earth 6032 times during her space flights on the International Space Station. If the International Space Station orbits the Earth 16 times per day, how many days was Petty Whitson in space?

Writing & Thinking

11. Explain how you would check a division problem that has a nonzero remainder.

12. Discuss how division is related to multiplication.

14.R.3 **Tests for Divisibility**

> ## Divisibility
>
> If a number can be divided by another number so that the remainder is 0, then we say
>
> 1. the number is _____
>
> 2. the divisor _____
>
> **DEFINITION**

> ## Divisibility by 2
>
> A number is divisible by 2 (is an **even number**) if _____
>
> **DEFINITION**

> ## Divisibility by 3
>
> A number is divisible by 3 if _____
>
> **DEFINITION**

▶ **Watch and Work**

Watch the video for Example 3 in the software and follow along in the space provided.

Example 2 **Determining Divisibility by 3**

Determine whether each of the following numbers is divisible by 3.

a. 6801

b. 356

Solution

✏ Now You Try It!

Use the space provided to work out the solution to the next example.

Example A Determining Divisibility by 3

Is 7912 divisible by 3? Explain why or why not.

Divisibility by 4

A number is divisible by 4 if _____

DEFINITION

Divisibility by 5

A number is divisible by 5 if _____

DEFINITION

Divisibility by 6

A number is divisible by 6 if _____

DEFINITION

Divisibility by 9

A number is divisible by 9 if _____

DEFINITION

Divisibility by 10

A number is divisible by 10 if _____

DEFINITION

14.R.3 Exercises

Concept Check

True/False. Determine whether each statement is true or false. If a statement is false, explain how it can be changed so the statement will be true. (**Note:** There may be more than one acceptable change.)

1. A number that is divisible by 10 is also divisible by 2 and 5.

2. 6801 is divisible by 9.

3. 7605 is divisible by 10.

4. 5,187,042 is divisible by 3.

Practice

Using the tests for divisibility, determine which of 2, 3, 4, 5, 6, 9, and 10 (if any) will divide exactly into each given number.

5. 105

6. 150

7. 331

8. 1234

Applications

Solve.

9. *Fundraising:* You are on a team that is participating in a charity walk with a goal to raise $12,400. Each team member agrees to raise the same amount of money. If the possible team sizes are 5, 6, 9, or 10 members, which team sizes allow the goal amount to be evenly split between the team members? How much money would each team member raise for each team size that can evenly split the goal amount?

10. *Time:* A company is working on a project that will take 440 hours of work to complete. The manager in charge of the project has the option to have 4, 6, or 8 people work on the project. If the manager wants to evenly divide the work between the team members, which team size will evenly split the work hours? How many hours would each team member spend on the project for each team size that evenly splits the work hours?

Writing & Thinking

11. **a.** If a number is divisible by both 3 and 5, then it will be divisible by 15. Give two examples.

    b. However, a number might be divisible by 3 and not by 5. Give two examples.

    c. Also, a number might be divisible by 5 and not 3. Give two examples.

14.R.4 Rules for Exponents

The Product Rule for Exponents

If a is a nonzero real number and m and n are integers, then

_____

In words, to multiply powers with the same base, _____

PROPERTIES

The Exponent 0

If a is a nonzero real number, then _____

The expression _____ is _____

DEFINITION

Quotient Rule for Exponents

If a is a nonzero real number and m and n are integers, then

_____

In words, to divide two powers with the same base, _____

PROPERTIES

Rule for Negative Exponents

If a is a nonzero real number and n is an integer, then

_____

PROPERTIES

▶ Watch and Work

Watch the video for Example 6 in the software and follow along in the space provided.

Example 6 Negative Exponents

Use the rule for negative exponents to simplify each expression so that it contains only positive exponents.

    a. 5^{-1}

    b. x^{-3}

    c. $x^{-9} \cdot x^{7}$

Solution

✏ Now You Try It!

Use the space provided to work out the solution to the next example.

Example A Negative Exponents

Use the rule for negative exponents to simplify each expression so that it contains only positive exponents.

a. 7^{-1}

b. x^{-7}

c. $x^{-11} \cdot x^{6}$

Summary of the Rules for Exponents

For any nonzero real number a and integers m and n:

1. The exponent 1: _____

2. The exponent 0: _____

3. The product rule:_____

4. The quotient rule:

_____

5. Negative exponents:

_____

PROPERTIES

14.R.4 **Exercises**

Concept Check

True/False. Determine whether each statement is true or false. If a statement is false, explain how it can be changed so the statement will be true. (**Note:** There may be more than one acceptable change.)

1. If a constant does not have an exponent written, it is assumed that the exponent is 0.

2. If a is a nonzero real number and n is an integer, then $a^{-n} = -a^n$.

3. Since the product rule is stated for integer exponents, the rule is also valid for 0 and negative exponents.

4. When using the quotient rule, you should subtract the smaller exponent from the larger exponent.

Practice

Simplify each expression. The final form of the expressions with variables should contain only positive exponents. Assume that all variables represent nonzero numbers.

5. $y^3 \cdot y^8$

6. $\dfrac{y^7}{y^2}$

7. $x^{-3} \cdot x^0 \cdot x^2$

8. $\dfrac{10^4 \cdot 10^{-3}}{10^{-2}}$

Applications

Solve.

9. *Computers:* Rylee wants to move all her files to a new hard drive that has 2^{12} GB of storage on it. She wants to designate the same amount of storage for each of 2^4 projects. How much storage should be assigned to each project? Write your answer as a power of two.

10. *Bacteria:* Trey is studying patterns in bacteria. For a positive test result in his experiment, bacteria must grow in population at a minimum rate of 3^2 in 24 hours. If the initial population of the bacteria is 3^5 and his final measurement after 24 hours is 3^8, should he mark the test as positive or negative?

14.R.5 Power Rules for Exponents

Power Rule for Exponents

If a is a nonzero real number and m and n are integers, then

\underline{\hspace{5cm}}

In other words, the value of a power raised to a power can be found by \underline{\hspace{6cm}}
\underline{\hspace{12cm}}

PROPERTIES

Rule for Power of a Product

If a and b are nonzero real numbers and n is an integer then

\underline{\hspace{5cm}}

In words, a power of a product is found by \underline{\hspace{6cm}}

PROPERTIES

Rule for Power of a Quotient

If a and b are nonzero real numbers and n is an integer, then

\underline{\hspace{5cm}}

In words, a power of a quotient (in fraction form) is found by \underline{\hspace{5cm}}
\underline{\hspace{12cm}}

PROPERTIES

▶ Watch and Work

Watch the video for Example 5 in the software and follow along in the space provided.

Example 5 Using Two Approaches with Fractional Expressions and Negative Exponents

Simplify: $\left(\dfrac{x^3}{y^5}\right)^{-4}$

Solution

✏️ Now You Try It!

Example A Using Two Approaches with Fractional Expressions and Negative Exponents

Simplify:

$$\left(\frac{x^6}{y^3}\right)^{-5}$$

Summary of the Rules for Exponents

For any nonzero real numbers a and b and integers m and n:

1. The exponent 1: _____

2. The exponent 0: _____

3. The product rule: _____

4. The quotient rule:
    _____

5. Negative exponents:
    _____

6. Power rule: _____

7. Power of a product: _____

8. Power of a quotient:
    _____

PROPERTIES

14.R.5 Exercises

Concept Check

True/False. Determine whether each statement is true or false. If a statement is false, explain how it can be changed so the statement will be true. (**Note:** There may be more than one acceptable change.)

1. Taking the reciprocal of a fraction changes the sign of any exponent in the fraction.

2. For an exponent to refer to −7 as the base, −7 must be in parentheses.

3. When simplifying an expression with exponents, the rules for exponents must be used in a specific order or the answer will vary.

4. The expression -8^2 simplifies to −64.

Practice

Use the rules for exponents to simplify each of the expressions. Assume that all variables represent nonzero real numbers.

5. $\left(2^{-3}\right)^{-2}$

6. $-3\left(7xy^2\right)^0$

7. $-2\left(3x^5y^{-2}\right)^{-3}$

8. $\left(\dfrac{x}{2}\right)^3$

9. $\left(\dfrac{2x^2y}{y^3}\right)^{-4}$

10. $\left(\dfrac{5a^4b^{-2}}{6a^{-4}b^3}\right)^{-2}\left(\dfrac{5a^3b^4}{2^{-2}a^{-2}b^{-2}}\right)^3$

14.R.6 Evaluating Radicals

Radical Terminology

The symbol $\sqrt{}$ is called _____

The number under the _____

The complete expression, such as $\sqrt{64}$ is called _____

DEFINITION

Square Root

If a is a nonnegative real number, then

_____

_____

_____

DEFINITION

Cube Root

If a is a real number, then _____

DEFINITION

▶ Watch and Work

Watch the video for Example 4 in the software and follow along in the space provided.

Example 4 Evaluating Cube Roots

a. Because $2^3 = \_$, $\sqrt[3]{8} = \_$.

b. Because $(-6)^3 = \_\_\_\_$, $\sqrt[3]{-216} = \_\_$.

_____

c. Because $\left(\dfrac{1}{3}\right)^3 = \_\_$, $\sqrt[3]{\dfrac{1}{27}} = \_$.

✎ Now You Try It!

Use the space provided to work out the solution to the next example.

Example A Evaluating Cube Roots

Evaluate the following radical expressions.

a. $\sqrt[3]{64}$

b. $\sqrt[3]{-125}$

c. $\sqrt[3]{\dfrac{1}{1000}}$

14.R.6 Exercises

Concept Check

True/False. Determine whether each statement is true or false. If a statement is false, explain how it can be changed so the statement will be true. (**Note:** There may be more than one acceptable change.)

1. If a number is squared and the principal square root of the result is found, that square root is always equal to the original number.

2. There is no real number that can be a square root of a negative number.

3. The index is the number underneath the radical sign.

4. The cube root of –27 is a real number.

Practice
Simplify the following square roots and cube roots.

5. $\sqrt{49}$

6. $\sqrt{289}$

7. $\sqrt[3]{1000}$

8. $\sqrt[3]{\dfrac{27}{64}}$

9. $\sqrt{0.04}$

Applications
Solve.

10. *Area:* The area of a square tile is 16 square inches.

    a. How long are the sides of the square tile?

    b. How many tiles would be needed for a four-foot-long and four-inch-high backsplash in a newly designed bathroom?

11. *Volume:* The volume of a child's building block is 64 cubic centimeters.

  a. Assuming the building block is a perfect cube, find the length of each side of the block.

  b. If a child stacks 5 blocks directly on top of each other, find the height of the structure that is created.

Writing & Thinking

12. Discuss, in your own words, why the square root of a negative number is not a real number.

13. Discuss, in your own words, why the cube root of a negative number is a negative number.

Math@Work

Math@Work

Basic Inventory Management

As an business manager you will need to evaluate the company's inventory several times per year. While evaluating the inventory, you will need to ensure that enough of each product will be in stock for future sales based on current inventory count, predicted sales, and product cost. Let's say that you check the inventory four times a year, or quarterly. You will be working with several people to get all of the information you need to make the proper decisions. You need the sales team to give you accurate predictions of how much product they expect to sell. You need the warehouse manager to keep an accurate count of how much of each product is currently in stock and how much of that stock has already been sold. You will also have to work with the product manufacturer to determine the cost to produce and ship the product to your company's warehouse. It's your job to look at this information, compare it, and decide what steps to take to make sure you have enough of each product in stock for sales needs. A wrong decision can potentially cost your company a lot of money.

Suppose you get the following reports: an inventory report of unsold products from the warehouse manager and the report on predicted sales for the next quarter (three months) from the sales team.

Unsold Products	
Item	Number in Stock
A	5025
B	150
C	975
D	2150

Predicted Sales	
Item	Expected Sales
A	4500
B	1625
C	1775
D	2000

Suppose the manufacturer gives you the following cost list for the production and shipment of different amounts of each inventory item.

Item	Amount	Cost	Amount	Cost	Amount	Cost
A	500	$875	1000	$1500	1500	$1875
B	500	$1500	1000	$2500	1500	$3375
C	500	$250	1000	$400	1500	$525
D	500	$2500	1000	$4250	1500	$5575

1. Which items and how much of each item do you need to purchase to make sure the inventory will cover the predicted sales?

2. If you purchase the amounts from Problem 1, how much will this cost the company?

3. By ordering the quantities you just calculated, you are ordering the minimum of each item to cover the expected sales. If the actual sales during the quarter are higher than expected, what might happen? How would you handle this situation?

4. Which skills covered in this chapter were necessary to help you make your decisions?

Math@Work

Hospitality Management: Preparing for a Dinner Service

As the manager of a restaurant you will need to make sure everything is in place for each meal service. This means that you need to predict and prepare for busy times, such as a Friday night dinner rush. To do this, you will need to obtain and analyze information to determine how much of each meal is typically ordered. After you estimate the number of meals that will be sold, you need to communicate to the chefs how much of each item they need to expect to prepare. An additional aspect of the job is to work with the kitchen staff to make sure you have enough ingredients in stock to last throughout the meal service.

You are given the following data, which is the sales records for the signature dishes during the previous four Friday night dinner services.

Week	Meal A	Meal B	Meal C	Meal D
1	30	42	28	20
2	35	38	30	26
3	32	34	26	26
4	30	32	28	22

Meal C is served with a risotto, a type of creamy rice. The chefs use the following recipe, which makes 6 servings of risotto, when they prepare Meal C. (**Note:** The abbreviation for tablespoon is T and the abbreviation for cup is c.)

$5\frac{1}{2}$ c chicken stock $\qquad$ $2\frac{1}{3}$ T chopped shallots $\qquad$ $\frac{1}{2}$ c red wine

$1\frac{1}{2}$ c rice $\qquad$ 2 T chopped parsley $\qquad$ $4\frac{3}{4}$ c thinly sliced mushrooms

2 T butter $\qquad$ 2 T olive oil $\qquad$ $\frac{1}{2}$ c Parmesan cheese

1. For the past four Friday night dinner services, what was the average number of each signature meal served? If the average isn't a whole number, explain why you would round this number either up or down.

2. Based on the average you obtained for Meal C, calculate how much of each ingredient your chefs will need to make the predicted amount of risotto.

3. The head chef reports the following partial inventory: $10\frac{3}{4}$ c rice, $15\frac{3}{4}$ c mushrooms, and 10 T shallots. Do you have enough of these three items in stock to prepare the predicted number of servings of risotto?

4. Which skills covered in this and the previous chapter helped you make your decisions?

Math@Work

Bookkeeper

As a bookkeeper, you will often receive bills and receipts for various purchases or expenses from employees of the company you work for. You will need to split the bill by expense code, assign costs according to customer, and reimburse an employee for their out-of-pocket spending. To do this you will need to know the company's reimbursement policies, the expense codes for different spending categories, and which costs fall into a particular expense category.

Suppose two employees from the sales department recently completed sales trips. Employee 1 flew out of state and visited two customers, Customer A and Customer B. This employee had a preapproved business meal with Customer B and was traveling for three days. Employee 2 drove out of state to visit Customer C. This employee stayed at a hotel for the night and then drove back the next day. The expenses for the two employees are as follows.

Employee 1	
Flight and Rental Car	$470.50
Hotel	$278.88
Meals	$110.56
Business Meal	$102.73
Presentation Materials	$54.86

Employee 2	
Miles Driven	578.5 miles
Fuel	$61.35
Hotel	$79.60
Meals	$53.23
Presentation Materials	$67.84

The expense categories used by your company to track spending are: Travel (includes hotel, flights, mileage, etc.), Meals (business), Meals (travel), and Supplies. Traveling employees are reimbursed up to $35 per day for meals while traveling and for all preapproved business meals. They also receive $0.565 per mile driven with their own car.

1. How much will you reimburse each employee for travel meals? Did either employee go over their allowed meal reimbursement amount?

2. What were the total expenses for each employee?

3. The company you work for keeps track of how much is spent on each customer. When a sales person visits multiple customers during one trip, the tracked costs are split between the customers. Fill in this table according to how much was spent on each customer for the different expense categories. (**Note:** For meals, only include the amount the employee was reimbursed.)

Expense	Customer A	Customer B	Customer C
Travel			
Meals (business)			
Meals (travel)			
Supplies			
Total			

Math@Work

Pediatric Nurse

As a pediatric nurse working in a hospital setting, you will be responsible for taking care of several patients during your work day. You will need to administer medications, set IVs, and check each patient's vital signs (such as temperature and blood pressure). While doctors prescribe the medications that nurses need to administer, it is important for nurses to double check the dosage amounts. Administering the incorrect amount of medication can be detrimental to the patient's health.

During your morning nursing round, you check in on three new male patients and obtain the following information.

	Patient A	Patient B	Patient C
Age	10	9	12
Weight (pounds)	81	68.5	112
Blood Pressure	97/58	100/59	116/73
Temperature (°F)	99.7	97.3	101.4
Medication	A	B	A

The following table shows the bottom of the range for abnormal blood pressure (BP) for boys. If either the numerator or the denominator of the blood pressure ratio is greater than or equal to the values in the chart, this can indicate a stage of hypertension.

Abnormal Blood Pressure for Boys by Age	
	Systolic BP / Diastolic BP
Age 9	109/72
Age 10	111/73
Age 11	113/74
Age 12	115/74
Source: http://www.nhlbi.nih.gov/health/public/heart/hbp/bp_child_pocket/bp_child_pocket.pdf	

Medication Directions	
Medication	Dosage Rate
A	40 mg per 10 pounds
B	55 mg per 10 pounds

1. Do any of the patients have a blood pressure which may indicate they have hypertension? If yes, which patient(s)?

2. Use proportions to determine the amount of medication that should be administered to each patient based on weight. Round to the nearest 10 pounds before calculating.

3. The average body temperature is 98.6 degrees Fahrenheit. You are supposed to alert the doctor on duty if any of the patients have a temperature 2.5 degrees higher than average. For which patients would you alert a doctor?

4. Which skills covered in this chapter and the previous chapters were necessary to help you make your decisions?

Math@Work

Architecture

As a project architect, you will be part of a team that creates detailed drawings of the project that will be used during the construction phase. It will be your job to ensure that the project will meet guidelines given to you by your company, such as square footage requirements and budget constraints. You will also need to meet the design requirements requested by the client.

Suppose you are part of a team that is designing an apartment building. You are given the task to create the floor plan for an apartment unit with two bedrooms and one bathroom. The apartment management company that has contracted your company to do the project has several requirements for this specific apartment unit.

1. One bedroom is the "master bedroom" and must have at least 60 square feet more than the other bedroom.

2. All walls must intersect or touch at 90 degree angles.

3. The kitchen must have an area of no more than 110 square feet.

4. The apartment must be between 1000 square feet and 1050 square feet.

A preliminary sketch of the apartment is shown here.

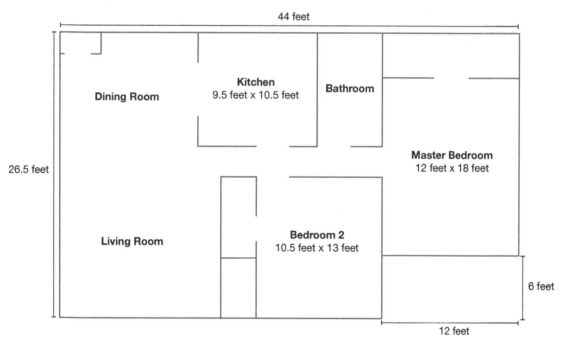

1. Does the apartment have the required total square footage that was requested? Is it over or under the total required?

2. Does the apartment blueprint meet the other requirements given by the client? If not, what does not meet the requirements?

3. For this specific apartment unit, the total construction cost per square foot is estimated to be $75.75. Approximately how much will it cost to construct each two-bedroom apartment based on the floor plan?

Math@Work

Statistician: Quality Control

Suppose you are a statistician working in the quality control department of a company that manufactures the hardware sold in kits to assemble book shelves, TV stands, and other ready-to-assemble furniture pieces. There are three machines that produce a particular screw and each machine is sampled every hour. A measurement of the screw length is determined with a micrometer, which is a device used to make highly precise measurements. The screw is supposed to be 3 inches in length and can vary from this measurement by no more than 0.1 inches or it will not fit properly into the furniture. The following table shows the screw length measurements (in inches) taken each hour from each machine throughout the day. The screw length data from each machine has also been plotted

Screw Length Measurements (in inches)			
Sample Time	Machine A	Machine B	Machine C
8 a.m.	2.98	2.92	2.99
9 a.m.	3.00	2.94	3.00
10 a.m.	3.02	2.97	3.01
11 a.m.	2.99	2.96	3.03
12 p.m.	3.01	2.94	3.05
1 p.m.	3.00	2.95	3.04
2 p.m.	2.97	2.93	3.06
3 p.m.	2.99	2.92	3.08
Mean			
Range			

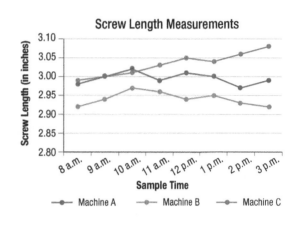

1. Calculate the mean and range of the data for each machine and place them in the bottom two rows of the table.

2. If the screw length can vary from 3 inches by no more than 0.1 inch (plus or minus), what are the lowest and highest values for length that will be acceptable? Place a horizontal line on the graph at each of these values on the vertical axis. These are the tolerance or specification limits for screw length.

3. Have any of the three machines produced an unacceptable part today? Are any of the machines close to making a bad part? If so, which one(s)?

4. Look at the graph and the means from the table that show the average screw length produced by each machine. Draw a bold horizontal line on the graph at 3 to emphasize the target length. Do all the machines appear to be making parts that vary randomly around the target of 3 inches?

5. Look at the range values from the table. Do any of the machines appear to have more variability in the length measurements than the others?

6. In your opinion, which machine is performing best? Would you recommend that any adjustments be made to any of the machines? If so, which one(s) and why?

Math@Work

Dental Assistant

As a dental assistant, your job duties will vary depending on where you work. Suppose you work in a dental office where you assist with dental procedures and managing patients' accounts. When a patient arrives for their appointment, you will need to review their chart and make sure they are up to date on preventive care, such as X-rays and cleanings. When the patient leaves, you will need to fill out an invoice to determine how much to charge the patient for their visit.

Dental patients generally have a new X-ray taken yearly. Cleanings are performed every 6 months, although some patients have their teeth cleaned more often. The following table shows the date of the last X-ray and cleaning for three patients that are visiting the office today. (**Note:** All dates are within the past year.)

Patient Histories		
Patient	**Last X–ray**	**Last Cleaning**
A	April 15	October 20
B	June 6	January 12
C	October 27	October 27

During Patient A's visit, she received a fluoride treatment and a cleaning. Patient A has no dental insurance. During Patient B's visit, he received a filling on one surface of a tooth. Patient B has dental insurance which pays for 60% of the cost of fillings. During Patient C's visit, he had a cleaning, a filling on one surface of a tooth, and a filling on two surfaces of another tooth. Patient C has dental insurance which covers the full cost of cleanings and 50% of the cost of fillings.

Fee Schedule	
Procedure	**Cost**
Cleaning	$95
Fluoride treatment	$35
Filling, One surface	$175
Filling, Two surfaces	$235
X–ray, Panoramic	$110

1. Using today's date, determine which of the three patients are due for a dental cleaning in the next two months?

2. Using today's date, determine which of the patients will require a new set of X-rays during this visit.

3. Determine the amount each patient will be charged for their visit (without insurance). Don't forget to include the cost of any X-rays that are due during the visit.

4. Use the insurance information to determine the amount that each patient will pay out-of-pocket at the end of their visit.

Math@Work

Financial Advisor

As a financial advisor working with a new client, you must first determine how much money your client has to invest. The client may have a lump sum that they have saved or inherited, or they may wish to contribute an amount monthly from their current salary. In the latter case, you must then have the client do a detailed budget, so that you can determine a reasonable amount that the client can afford to set aside on a monthly basis for investment.

The second piece of information necessary when dealing with a new client is determining how much risk-tolerance they have. If the client is young or has a lot of money to invest, they may be willing to take more risk and invest in more aggressive, higher interest-earning funds. If the client is older and close to retirement, or has little money to invest, they may prefer less-aggressive investments where they are essentially guaranteed a certain rate of return. The range of possible investments that would suit each client's needs and goals are determined using a survey of risk-tolerance.

Suppose you have a client who has a total of $25,000 to invest. You determine that there are two investment funds that meet the client's investment preferences. One option is an aggressive fund that earns an average of 12% interest and the other is a more moderate fund that earns an average of 5% interest. The client desires to earn $2300 this year from these investments.

Investment Type	Principal Invested	·	Interest Rate	=	Interest Earned
Aggressive Fund	x				
Moderate Fund					

To determine the amount of interest earned you know to use the table above and the formula $I = Prt$, where I is the interest earned, P is the principal or amount invested, r is the average rate of return, and t is the length of time invested. Since the initial investment will last one year, $t = 1$.

1. Fill in the table with the known information. If x is the amount invested in the aggressive fund and the total amount to be invested is $25,000, create an expression involving x for the amount that will be left to invest in the moderate fund. Place this expression in the appropriate cell of the table.

2. Determine an expression in x for the interest earned on each investment type by multiplying the principal by the interest rate.

3. Determine the amount invested in each fund by setting up an equation using the expressions in column four and the fact that the client desires to earn $2300 from the interest earned on both investments.

4. Verify that the investment amounts calculated for each fund in the previous step are correct by calculating the actual interest earned in a year for each and making sure they sum to $2300.

5. Why would you not advise your client to invest all their money in the fund earning 12% interest, after all, it has the highest average interest rate?

Math@Work

Market Research Analyst

As a market research analyst, you may work alone at a computer, collecting and analyzing data, and preparing reports. You may also work as part of a team or work directly with the public to collect information and data. Either way, a market research analyst must have strong math and analytical skills and be very detail-oriented. They must have strong critical-thinking skills to assess large amounts of information and be able to develop a marketing strategy for the company. They must also possess good communication skills in order to interpret their research findings and be able to present their results to clients.

Suppose you work for a shoe manufacturer who wants to produce a new type of lightweight basketball sneaker similar to a product a competitor recently released into the market. You have gathered some sales data on the competitor in order to determine if this venture would be worthwhile, which is shown in the table below. To begin your analysis, you create a scatter plot of the data to see the sales trend. (A scatter plot is a graph made by plotting ordered pairs in a coordinate plane in order to show the relationship between two variables.) You determine that the x-axis will represent the number of weeks after the competitors new sneaker went on the market and the y-axis will represent the amount of sales in thousands of dollars.

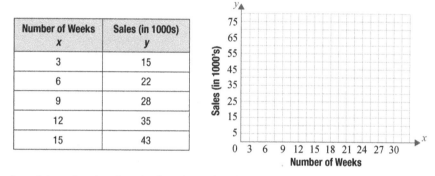

Number of Weeks x	Sales (in 1000s) y
3	15
6	22
9	28
12	35
15	43

1. Create a scatter plot of the sales data by plotting the ordered pairs in the table on the coordinate plane. Does the data on the graph appear to follow a linear pattern? If so, sketch a line that you feel would "best" fit this set of data. (A market research analyst would typically use computer software to perform a technique called regression analysis to fit a "best" line to this data.)

2. Using the ordered pairs corresponding to weeks 9 and 15, find the equation of a line running through these two data points.

3. Interpret the value calculated for the slope of the equation in Problem 2 as a rate of change in the context of the problem. Write a complete sentence.

4. If you assume that the sales trend in sneaker sales follows the model determined by the linear equation in Problem 2, predict the sneaker sales in 6 months. Use the approximation that 1 month is equal to 4 weeks.

5. Give at least two reasons why the assumption made in Problem 4 may be invalid?

Math@Work

Chemistry

As a pharmaceutical chemist, you will need an advanced degree in pharmaceutical chemistry, which combines biology, biochemistry, and pharmaceuticals. In this career, you will most likely spend your day in a lab setting creating new medications or researching their effectiveness. You will often work as part of a team working towards a joint goal. As a result, in addition to strong math skills and an understanding of chemistry, you will need to have good communication and leadership skills. Since you will be working directly with chemicals, you will also need to have a strong understanding of lab safety rules to ensure the safety of not only yourself, but your coworkers as well.

Suppose you work at a pharmaceutical company which creates and produces medications for various skin conditions. You are currently on a team which is developing an acne-controlling facial cleanser. Your team is working on determining the gentlest formula possible that is still effective so that the cleanser can be used on sensitive skin. Half of your team is working with salicylic acid and the other half is working with benzoyl peroxide.

As a part of your work, you will need to keep up on current research. Learning about new chemicals, new methods, and new research will be a continuous part of your life.

1.   Perform an Internet search for benzoyl peroxide. How does it work to clean skin and prevent acne?

2.   Perform an Internet search for salicylic acid. How does it work to clean skin and prevent acne?

3.   Based on your research, which chemical seems better suited to treat acne on sensitive skin?

Another aspect of your career will involve the mixing of chemicals to create new compounds. Having the correct concentrations of chemicals is also important so the resulting solution works as you expect it to. When you don't have the correct concentration of a chemical in stock, it is possible to mix two concentrations together to obtain the desired concentration.

4.   Your team wants to create a cleanser with 4% benzoyl peroxide. The lab currently has 2.5% and 10% concentrations of benzoyl peroxide in stock. To create 500 mL of 4% benzoyl peroxide, how much of each concentration should be combined?

Math@Work

Astronomy

Astronomy is the study of celestial bodies, such as planets, asteroids, and stars. While you work in the field of astronomy, you will use knowledge and skills from several other fields, such as mathematics, physics, and chemistry. An important tool of astronomers is the telescope. Several powerful telescopes are housed in observatories around the world. One of the many things astronomers use observatories for is discovering new celestial objects such as a near-Earth object (NEO). NEOs are comets, asteroids, and meteoroids that orbit the sun and cross the orbital path of Earth. The danger presented by NEOs is that they may strike the Earth and result in global catastrophic damage. (**Note:** The National Aeronautics and Space Administration (NASA) keeps track of all NEOs which are a potential threat at the website http://neo.jpl.nasa.gov/risk/)

For an asteroid to be classified as an NEO, the asteroid must have an orbit that partially lies within 0.983 and 1.3 astronomical units (AU) from the sun, where 1 AU is the furthest distance from the Earth to the sun, approximately 9.3×10^7 miles.

Near-Earth Object Distance			
	Minimum		Maximum
Distance in AU	0.983 AU	1 AU	1.3 AU
Distance in Miles		9.3×10^7 miles	

Suppose you discover three asteroids that you suspect may be NEOs. You perform some calculations and come up with the following facts. The furthest that Asteroid A is ever from the sun is 81,958,000 miles. The closest Asteroid B is ever to the sun is 12,529,000 miles. The closest Asteroid C is ever to the sun is 92,595,000 miles.

1. To determine if any of the asteroids pass within the range to be classified as an NEO, fill in the missing values from the table.

2. Based on the measurements from Problem 1, do any of the three asteroids qualify as an NEO?

There are two scales that astronomers use to explain the potential danger of NEOs. The Torino Scale is a scale from 0 to 10 that indicates the chance that an object will collide with the Earth. A rating of 0 means there is an extremely small chance of a collision and a 10 indicates that a collision is certain to happen. The Palermo Technical Impact Hazard Scale is used to rate the potential impact hazard of an NEO. If the rating is less than −2, the object poses a very minor threat with no drastic consequences if the object hits the Earth. If the rating is between −2 and 0, then the object should be closely monitored as it could cause serious damage.

Go to the NASA website http://neo.jpl.nasa.gov/risk/ to answer the following questions.

3. Does any NEO have a Torino Scale rating higher than 0? If so, what is the object's designation (or name) and during which year range could a potential impact occur?

4. Which NEO has the highest Palermo Scale rating? During which year range could a potential impact occur?

Math@Work

Math Education

As a math instructor at a public high school, your day will be spent preparing class lectures, grading assignments and tests, and teaching students with a wide variety of backgrounds. While teaching math, it is your job to explain the concepts and skills of math in a variety of ways to help students learn and understand the material. As a result, a solid understanding of math and strong communication skills are very important. Teaching math is a challenge and being able to understand the reasons that students struggle with math and empathize with these students is a critical aspect of the job.

Suppose that the next topics you plan to teach to your algebra students involve finding the greatest common factor and factoring by grouping. To teach these skills, you will need to plan how much material to cover each day, choose examples to walk through during the lecture, and assign in-class work and homework. You decide to spend the first day on this topic explaining how to find the greatest common factor of a list of integers.

1. It is usually easier to teach a group of students a new topic by initially showing them a single method. If a student has difficulty with that method, then showing the student an alternative method can be helpful. Which method for finding the greatest common factor would you teach to the class during the class lecture?

2. On a separate piece of paper, sketch out a short lecture on finding the greatest common factor of a list of integers. Be sure to include examples that range from easy to difficult.

3. While the class is working on an in-class assignment, you find that a student is having trouble following the method that you taught to the entire class. Describe an alternative method that you could show the student.

4. From your experience with learning how to find the greatest common factor of a list of integers, what do you think are some areas that might confuse students and cause them to struggle while learning this topic? Explain how understanding the areas that might cause confusion can help you become a better teacher.

Math@Work

Physics

As an employee of a company that creates circuit boards, your job may vary from designing new circuit boards, setting up machines to mass produce the circuit boards, to testing the finished circuit boards as part of quality control. Depending on your position, you may work alone or as part of a team. Regardless of who you work with, you will need strong math skills to be able to create new circuit board designs and strong communication skills to describe the specifications for a new circuit board design, describe how to set up the production line, or explain why a part is faulty.

Suppose your job requires you to create new circuit boards for a variety of electronic equipment. The latest circuit board that you are designing is a small part of a complicated device. The circuit board you create has three resistors which run in parallel, as shown in the diagram.

Two of the resistors were properly labeled with their correct resistance, which is measured in ohms. The first resistor has a rating of 2 ohms. The second resistor has a rating of 3 ohms. The third resistor was taken from the supply shelf for resistors of a certain rating, but the resistor was unlabeled. As a result, you are unsure if it has the correct resistance for the current you want to produce. You use an ohmmeter, a device that measures resistance in a circuit, to determine that the total resistance of the circuit you created is $\frac{30}{31}$ ohms.

You know that the equation to determine the total resistance R_t is $\frac{1}{R_t} = \frac{1}{R_1} + \frac{1}{R_2} + \frac{1}{R_3}$, where R_1 is the resistance of the first resistor, R_2 is the resistance of the second resistor, and R_3 is the resistance of the third resistor.

1. Use the formula to determine the resistance of the third resistor given that the total resistance of the circuit is $\frac{30}{31}$ ohms.

2. Was the third resistor on the correct shelf if you took it from the supply shelf that holds resistors with a rating of 7 ohms?

3. What would be the total resistance of the circuit if the third resistor had a rating of 7 ohms?

4. What do you think would happen if the resistance of the unlabeled resistor wasn't determined and the circuit board was sent to the production line to be mass produced?

Math@Work

Forensic Scientist

As a forensic scientist, you will work as part of a team to investigate the evidence from a crime scene. Every case you encounter will be unique and the work may be intense. Communication is especially important because you will need to be clear and honest about your findings and your conclusions. A suspect's freedom may depend on the conclusions your team draws from the evidence.

Suppose the most recent case that you are involved in is a hit-and-run accident. A body was found at the side of the road with skid marks nearby. The police are unsure if the cause of death of the victim was vehicular homicide. Among the case description, the following information is provided to you.

Accident Report	
Date:	June 14
Time:	9:30 pm
Climate:	55 degrees Fahrenheit, partly cloudy, dry
Description of crime scene:	
Victim was found at the side of a road. Body temperature upon arrival is 84.9 °F. Posted speed limit is 30 mph. Road is concrete. Conditions are dry. Skid marks near the body are 88 feet in length.	

Known formulas and data:

A body will cool at a rate of 2.7 °F per hour until the body temperature matches the temperature of the environment. Average human body temperature is 98.6 °F.

Impact Speed and Risk of Death	
Impact Speed	**Risk of Death**
23 mph	10%
32 mph	25%
42 mph	50%
58 mph	90%
Source: 2011 AAA Foundation for Traffic Safety "Impact Speed and Pedestrian's Risk of Severe Injury or Death"	

Braking distance is calculated using the formula $\dfrac{s}{\sqrt{l}} = k$, where s is the initial speed of the vehicle in mph, l is the length of the skid marks in feet, and k is a constant that depends on driving conditions. Based on the driving conditions on that road for the last 12 hours, $k = \sqrt{20}$.

1. Based on the length of the skid marks, how fast was the car traveling before it attempted to stop? Round to the nearest whole number.

2. Based on the table, what percent of pedestrians die after being hit by a car moving at that speed?

3. Based on the cooling of the body, if the victim died instantly, how long ago did the accident occur? Round to the nearest hour.

4. Can you think of any other factors that should be taken into consideration before determining whether the impact of the car was the cause of death?

Math@Work

Other Careers in Mathematics

Earning a degree in mathematics or minoring in mathematics can open many career pathways. While a degree in mathematics or a field which uses a lot of mathematics may seem like a difficult path, it is something anyone can achieve with practice, patience, and persistence. Three growing fields of study which rely on mathematics are actuarial science, computer science, and operations research. While each of these fields involves mathematics, they require special training or additional education outside of a math degree. A brief description of each career is provided below along with a source to find more information about these careers.

Growing Fields of Study

Actuarial Science: The field of actuarial science uses methods of mathematics and statistics to evaluate risk in industries such as finance and insurance. Visit www.beanactuary.org for more information

Computer Science: From creating web pages and computer programs to designing artificial intelligence, computer science uses a variety of mathematics. Visit www.acm.org for more information.

Operations Research: The discipline of operations research uses techniques from mathematical modeling, statistical analysis, and mathematical optimization to make better decisions, such as maximizing revenue or minimizing costs for a business. Visit www.informs.org for more information.

There are numerous careers that have not been discussed in this workbook. Exploring career options before choosing a major is a very important step in your academic career. Learning about the career you are interested in before completing your degree can help you choose courses that will align with your career goals. You should also explore the availability of jobs in your chosen career and whether you will have to relocate to another area to be hired. The following web sites will help you find information related to different careers that use mathematics. Another great resource is the mathematics department at your college.

The **Mathematical Association of America** has a website with information about several careers in mathematics. Visit www.maa.org/careers to learn more.

The **Society for Industrial and Applied Mathematics** also has a webpage dedicated to careers in mathematics. Visit www.siam.org/careers to learn more.

The **Occupational Outlook Handbook** is a good source for information on educational requirements, salary ranges, and employability of many careers, not just those that involve mathematics. Visit http://www.bls.gov/ooh/ to learn more.

Answer Key

Chapter 1.R: Critical Thinking and Problem Solving

1.R.1 Exercises

Concept Check

1. True

3. True

Practice

5. 2: ten thousands, 4: thousands, 6: hundreds, 8: ones

7. Six hundred eighty-three thousand, one hundred

Applications

9. Eighty-two thousand, one hundred three

Writing & Thinking

11. 0 is a whole number and not a natural number. Both whole numbers and natural numbers include 1, 2, 3, 4, 5, and so on.

1.R.2 Exercises

Concept Check

1. True

3. True

Practice

5. 220; 223

7. 40,000; 43,680

Applications

9. $40,000; $35,316

Writing & Thinking

11. Estimation uses rounded values to find an approximate sum, difference, product, etc. Answers will vary.

1.R.3 Exercises

Concept Check

1. False; Equals 81

3. False; 7^0 is 1.

Practice

5. **a.** 2 **b.** 3 **c.** 8

7. 2

Applications

9. **a.** No. Here it shows that we are only dividing the old trading cards by 6 friends versus both the old and new trading cards by 6 friends.

   b. 522; $\dfrac{15 \cdot 10 \cdot 20 + 132}{6}$

Writing & Thinking

11. If addition is within parentheses (or other grouping symbols), addition would be performed first.

1.R.4 Exercises

Concept Check

1. True

3. False; Quotient indicates division.

Applications

5. 1103 calories

7. 380 sq in.

Writing & Thinking

9. Answers will vary.

1.R.5 Exercises

Concept Check

1. True

3. False; Subtraction is indicated by the phrase "five less than a number."

Practice

5. $x + 6$

7. $\dfrac{x}{2} - 18$

9. **a.** $4n - 6$

   b. $6 - 4n$

11. The product of a number and negative nine

Writing & Thinking

13. The Commutative Property of Addition and Multiplication permits the order of items being added or multiplied to change and still have the same result. This property does not hold true for subtraction or division. Therefore, order is impor-

tant for subtraction and division problems or the answer will change or be incorrect.

1.R.6 Exercises

Concept Check

1. False; The addition and multiplication principles of equality can be used with decimal or fractional coefficients.

3. True

Practice

5. $x = -3$

7. $x = -\dfrac{27}{10}$

Applications

9. 14,000 tickets per hour

Writing & Thinking

11. **a.** The 4 should have been multiplied by 3 so that the 3 was distributed over the entire left-hand side of the equation; Correct answer is $x = 15$.

   b. 3 should be subtracted from each side, not from each term, and $5x - 3$ doesn't simplify to $2x$; Correct answer is $x = \dfrac{8}{5}$.

Chapter 2.R: Set Theory

2.R.1 Exercises

Concept Check

1. True

3. True

5.

7. 0, 4, 8

9. True

Applications

11. −4500 meters

Writing & Thinking

13. If y is a negative number then $-y$ represents a

positive number. For example, if $y = -2$, then $-y = -(-2) = 2$.

2.R.2 Exercises

Concept Check

1. False; The sum of a positive and negative number can be positive, negative, or zero.

3. False; The sum of two positive numbers is always positive, zero is neither positive nor negative.

Practice

5. −6

7. −0.5

Applications

9. **a.** $45,000 + (−$8000) + (−$2000) + $15,000.

   b. $50,000.

Writing & Thinking

11. $|0| + |0| = 0$

2.R.3 Exercises

Concept Check

1. False; The sum of a number and its additive inverse is zero.

3. True

Practice

5. −11

7. 3

9. −15

11. −18°F (a decrease of 18 degrees Fahrenheit)

Writing & Thinking

13. Add the opposite of the second number to the first number.

2.R.4 Exercises

Concept Check

1. True

3. False; The product and quotient will be positive.

Practice

5. 48

7. 2

Applications

9. −12

Writing & Thinking

11. Negative; The product of every two negative numbers will be positive and this result multiplied by the remaining negative will give a negative answer.

2.R.5 Exercises

Concept Check

1. True

3. True

Practice

5. **a.** 36

   b. 16

7. −10

9. 129

Applications

11. **a.** −$42 − $35 − (3 · $5).

    b. −$92.

Writing & Thinking

13. Smaller; When any positive number is multiplied by a fraction (or decimal) between 0 and 1, the result will be smaller. This is what is happening when a number between 0 and 1 is squared. Answers will vary.

Chapter 4.R: Rates, Ratios, Proportions, and Percentages

4.R.1 Exercises

Concept Check

1. False; The numerator is 11.

3. True

Practice

5.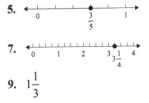

7. ![number line from 0 to 4 with point at 3 1/4]
   $0 \quad 1 \quad 2 \quad 3 \ 3\frac{1}{4} \ 4$

9. $1\frac{1}{3}$

Applications

11. $\dfrac{115}{146}$

Writing & Thinking

13. Multiply the denominator by the whole number and add the numerator. This number is the new numerator. The denominator stays the same.

4.R.2 Exercises

Concept Check

1. True

3. False; On a number line, any number to the right of another number is larger than that other number.

Practice

5. 2.57

7. 6.028

9. **a.** 5

   b. 2

   c. 2, 5, 2

   d. 3.0065

Applications

11. Two and eight-hundred twenty-five ten-thousandths

Writing & Thinking

13. Moving left to right, compare digits with the same place value. When one compared digit is larger, the corresponding number is larger.

4.R.3 Exercises

Concept Check

1. True

3. False; A decimal number that is between 0.01 and 0.10 is between 1% and 10%.

Practice

5. 20%

7. 2%

9. 0.07

Applications

11. 4%

Writing & Thinking

13. Percent means per centum or per 100. For example, fifty-eight percent means 58 out of 100. Percent can be written as a fraction with 100 in the denominator as in 58/100. The decimal equivalent, 0.58 is read as "fifty-eight hundredths," indicating percent can be written using the hundredths place, another connection.

4.R.4 Exercises

Concept Check

1. False; Fractions that have denominators other than 100 can be changed to a percent.

3. False; When changing from a percent to a mixed number, the fraction should be reduced.

Practice

  5. 75%

  7. $1\frac{1}{5}$

Applications

  9. 85%

Writing & Thinking

  11. 100% = 1 so anytime there is a mixed number, which has a value greater than 1, the percentage will be greater than 100%. Proper fractions (numerator is smaller than denominator) have a value less than 1 and therefore the percentage will be less than 100%.

4.R.5 Exercises

Concept Check

  1. True

  3. False; In the problem "What is 26% of 720?" the missing number is the amount.

Practice

  5. 7.5

  7. 15

  9. 250

Applications

  11. $97,600

Writing & Thinking

  13. Proportions would work for mixed numbers because a mixed number can be rewritten as a fraction. The only additional step required would be to change the mixed number to an improper fraction and then solve the proportion as normal.

Chapter 5.R: The Mathematics of Growth

5.R.1 Exercises

Concept Check

  1. True

  3. True

Practice

  5. $\begin{cases} A(-5,1),\ B(-3,3), \\ C(-1,1),\ D(1,2), \\ E(2,-2) \end{cases}$

  7. a. $(0,-1)$

    b. $(4,1)$

    c. $(2,0)$

    d. $(8,3)$

  9. b, c

  11. a.

D	E
100	85
200	170
300	255
400	340
500	425

    b.

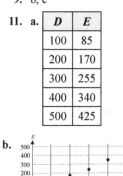

5.R.2 Exercises

Concept Check

  1. True

  3. False; Horizontal lines have y-intercepts

Practice

  5.

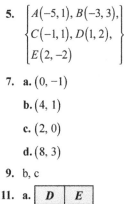

  7.

  9.

  11. The y-intercept is $(0,30)$, meaning that if a student does no homework at all, the student will get a score of 30 points on the exam.

  13. Substitute the x and y values into the equation. Then evaluate both sides to see if the equation is true.

5.R.3 Exercises

Concept Check

  1. False; If there is no exponent written, the exponent is assumed to be 1.

  3. True

Practice

  5. y^{11}

  7. $\dfrac{1}{x}$

Application

  9. 2^8 GBs

5.R.4 Exercises

Concept Check

  1. False; Variables need to be considered as well.

Practice

  3. 2

  5. $13ab$

Applications

  7. She can make 30 identical treat bags, each containing 5 pieces of candy A, 6 pieces of candy B, and 11 pieces of candy C.

5.R.5 Exercises

Concept Check

  1. True

  3. True

Practice

  5. $(x-9)(x+3)$

  7. $(x-12)(x-2)$

Application

  9. Base $= x+48$; Height $= x$

Writing & Thinking

  11. If the sign of the constant term is positive, the signs in the factors will both be positive or both be negative. If the sign of the constant term is negative, the sign in one factor will be positive and the sign in the other factor will be negative.

5.R.6 Exercises

Concept Check

  1. False; The middle term should be the sum of the inner and outer products.

  3. True

Practice

5. $-(x-2)(x-1)$

7. $(3x+4)(3x-5)$

9. $(5a-2)(a-1)$

Writing & Thinking

11. This is not the completely factored form because $(5x-15)$ can be factored into $5(x-3)$.

5.R.7 Exercises

Concept Check

1. True

3. False; The sum of two squares is not factorable.

Practice

5. $(y-8)^2$

7. **a.** x^2-16

b.
$$\boxed{}\ x-4$$
$$x+4$$

5.R.8 Exercises

Concept Check

1. True

3. False; Two real solutions

Practice

5. 68; Two real solutions

7. $x=-2\pm2\sqrt{2}$

9. $x=1,\dfrac{4}{3}$

Writing & Thinking

11. $x^4-13x^2+36=0$; multiplied $(x-2)(x+2)(x-3)(x+3)$

Chapter 6.R: Geometry

6.R.1 Exercises

Concept Check

1. False; A proportion is a statement that two ratios are equal.

3. True

5. True

Practice

7. False

9. $B=7.8$

Applications

11. 180 minutes or 3 hours

6.R.2 Exercises

Concept Check

1. True

3. True

Practice

5. 6

7. Yes, $62+82=102$

9. $c=5$

Applications

11. 17.0 inches

6.R.3 Exercises

Concept Check

1. True.

Practice

3. $15x$

5. $10x^2+28x-76$

7. $8a$

6.R.4 Exercises

Concept Check

1. True.

Practice

3. $3x+4$; 13

Applications

5. $50,000

Writing & Thinking

7. Like terms have the same variables with the same exponents. For example, $4a^2bc^3$ and $-3a^2bc^3$ are like terms. Unlike terms either have different variables or possibly the same variables with different exponents. For example

$6ab$ and $-9a^2b$ are unlike terms and $5xy$ and $13ax$ are unlike terms.

6.R.5 Exercises

Concept Check

1. False; Case matters in formulas.

3. True

Applications

5. $1030

7. $b=P-a-c$

9. $t=\dfrac{I}{Pr}$

Chapter 7.R: Probability

7.R.1 Exercises

Concept Check

1. True

3. True

5. False; The product of a nonzero number and its reciprocal is 1.

Practice

7. $\dfrac{1}{3}$

9. 0

10. Undefined

11. $\dfrac{10}{9}$

13. 4

Applications

15. $\dfrac{3}{8}$

17. 200 years

19. **a.** $22\dfrac{1}{2}$ gallons

b. $45

Writing & Thinking

21. To multiply two fractions, multiply the numerators, multiply the denominators, and then reduce the product to lowest terms. Examples will vary.

23. No. For example, $\dfrac{4}{5}\neq\dfrac{5}{4}$. Answers will vary.

7.R.2 Exercises

Concept Check

1. False; The LCM of 15 and 25 is 75.

3. False; The first five multiples of 4 are 4, 8, 12, 16, and 20.

Practice

5. 30

7. **a.** LCM = 490

b. $490 = 14 \cdot 35$
$= 35 \cdot 14 = 49 \cdot 10$

9. 45

Applications

11. a. 360 pieces

b. 15 boxes, 10 boxes, and 8 boxes, respectively

Writing & Thinking

13. Multiplying the two numbers together will give the LCM if those two numbers have no common factors. If they have any factors in common, then you would only use that

common factor once. Examples will vary.

7.R.3 Exercises

Concept Check

1. True

3. False; LCD stands for least common denominator.

5. True

Practice

7. $\dfrac{17}{21}$

9. $\dfrac{23}{42}$

Applications

11. 1 ounce

13. The LCM finds the least common multiple of a set of numbers. The LCD does the same thing for the set of numbers determined by the denominators.

7.R.4 Exercises

Concept Check

1. True

3. False; In some cases, fractions can be converted to decimal form without losing accuracy.

Practice

5. $\dfrac{9}{50}$

7. 6.67

9. 0.7

Applications

11. 17.92 inches

Writing & Thinking

13. For the numerator, write the whole number formed by all the digits of the decimal number, and for the denominator, write the power of 10 that corresponds to the rightmost digit. Reduce the fraction, if possible.

Chapter 8.R: Statistics

8.R.1 Exercises

Concept Check

1. True

3. False; A decimal number that is between 0.01 and 0.10 is between 1% and 10%.

Practice

5. 20%

7. 2%

9. 0.07

Applications

11. 4%

Writing & Thinking

13. Percent means per centum or per 100. For example, fifty-eight percent means 58 out of 100. Percent can be written as a fraction with 100 in the denominator as in 58/100. The decimal equivalent, 0.58 is read as "fifty-eight hundredths," indicating percent can be written using the hundredths place, another connection.

8.R.2 Exercises

Concept Check

1. False; Fractions that have denominators other than 100 can be changed to a percent.

3. False; When changing from a percent to a mixed number, the fraction should be reduced.

Practice

5. 75%

7. $1\dfrac{1}{5}$

Applications

9. 85%

Writing & Thinking

11. 100% = 1 so anytime there is a mixed number, which has a value greater than 1, the percentage will be greater than 100%. Proper fractions (numerator is smaller than denominator) have a value less than 1 and therefore the percentage will be less than 100%.

8.R.3 Exercises

Concept Check

1. False; Case matters in formulas.

3. True

Applications

5. $1030

7. $b = P - a - c$

9. $t = \dfrac{I}{Pr}$

8.R.4 Exercises

Concept Check

1. True

3. True

Practice

5. $\begin{cases} A(-5, 1),\ B(-3, 3), \\ C(-1, 1),\ D(1, 2), \\ E(2, -2) \end{cases}$

7. a. $(0, -1)$

b. $(4, 1)$

c. $(2, 0)$

d. $(8, 3)$

9. b, c

11. a.

D	E
100	85
200	170
300	255
400	340
500	425

b.

8.R.5 Exercises

Concept Check

1. True

3. False; Horizontal lines have y-intercepts

Practice

5.

7.

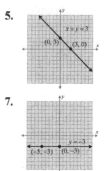

9.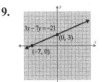

11. The y-intercept is $(0, 30)$, meaning that if a student does no homework at all, the student will get a score of 30 points on the exam.

13. Substitute the x and y values into the equation. Then evaluate both sides to see if the equation is true.

8.R.6 Exercises

Practice

1. m is undefined

3.

Vertical line; m is undefined

5.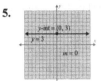

7. \$4000/year

9. a. For any horizontal line, all of the y values

will be the same. Thus the formula for slope will always have 0 in the numerator making the slope of every horizontal line 0.

b. For any vertical line, all of the x values will be the same. Thus the formula for slope will always have 0 in the denominator making the slope of every vertical line undefined.

8.R.7 Exercises

Concept Check

1. False; If the original number is negative, the principal square root will

not be the same as the original number.

3. False; The radicand is underneath the radical symbol.

Practice

5. 7

7. 10

9. 0.2

Application

11. a. 4 cm

b. 20 cm

Writing & Thinking

13. Cubing a negative real number is equivalent to multiplying a negative number by itself 3 times. The product of three negative numbers is negative.

Chapter 9.R: Personal Finance

9.R.1 Exercises

Concept Check

1. True

3. True

Practice

5. 2: ten thousands, 4: thousands, 6: hundreds, 8: ones

7. Six hundred eighty-three thousand, one hundred

Applications

9. Eighty-two thousand, one hundred three

Writing & Thinking

11. 0 is a whole number and not a natural number. Both whole numbers and natural numbers include 1, 2, 3, 4, 5, and so on.

9.R.2 Exercises

Concept Check

1. False; A polygon has three or more sides.

3. False; Borrowing must occur.

Practice

5. 58

7. 144

9. 42 cm

Applications

11. \$39,100

Writing & Thinking

13. Answers will vary. Subtraction may be used when paying bills, buying items, losing weight, etc.

9.R.3 Exercises

Concept Check

1. False; Equals 81

3. False; 7^0 is 1.

Practice

5. a. 2 **b.** 3 **c.** 8

7. 2

Applications

9. a. No. Here it shows that we are only dividing the old trading cards by 6 friends versus both the old and new trading cards by 6 friends.

b. 522; $\dfrac{15 \cdot 10 \cdot 20 + 132}{6}$

Writing & Thinking

11. If addition is within parentheses (or other grouping symbols), addition would be performed first.

9.R.4 Exercises

Concept Check

1. True

3. False; On a number line, any number to the right of another number is larger than that other number.

Practice

5. 2.57

7. 6.028

9. a. 5

b. 2

c. 2, 5, 2

d. 3.0065

Applications

11. Two and eight-hundred twenty-five ten-thousandths

Writing & Thinking

13. Moving left to right, compare digits with the same place value. When one compared digit is larger, the corresponding number is larger.

9.R.5 Exercises

Concept Check

1. True

3. False; A decimal number that is between 0.01 and 0.10 is between 1% and 10%.

Practice

5. 20%

7. 2%

9. 0.07

Applications

11. 4%

Writing & Thinking

13. Percent means per centum or per 100. For example, fifty-eight percent means 58 out of 100. Percent can be written as a fraction with 100 in the denomi-

nator as in 58/100. The decimal equivalent, 0.58 is read as "fifty-eight hundredths," indicating percent can be written using the hundredths place, another connection.

9.R.6 Exercises

Concept Check

1. False; In order to solve the equation $0.56 \cdot B = 12$ for the base, B, one would divide 12 by 0.56.

3. False; The solution to the problem "50% of what number is 352?" could be found by solving the equation $0.5 \cdot B = 352$.

Practice

5. 7

7. 42

9. 20

Applications

11. $175,000

13. The amount is the number that is often near the word "is." The base is the number that often follows the word "of." The rate is the number written either as a fraction or as a decimal number that has not been identified as the amount or the base, and usually appears before the word "of."

9.R.7 Exercises

Concept Check

1. True

3. False; In the term "12a," 12 is the coefficient.

Practice

5. −5, 3, and 8 are like terms; 7x and 9x are like terms.

7. 10x

9. $3x + 4$; 13

Applications

11. $50,000

Writing & Thinking

13. Like terms have the same variables with the same exponents. For example, $4a^2bc^3$ and $-3a^2bc^3$ are like terms. Unlike terms either have different variables or possibly the same variables with different exponents. For example $6ab$ and $-9a^2b$ are unlike terms and $5xy$ and $13ax$ are unlike terms.

Chapter 10.R: Voting and Apportionment

10.R.1 Exercises

Concept Check

1. False; A polygon has three or more sides.

3. False; Borrowing must occur.

Practice

5. 58

7. 144

9. 42 cm

Applications

11. $39,100

Writing & Thinking

13. Answers will vary. Subtraction may be used when paying bills, buying items, losing weight, etc.

10.R.2 Exercises

Concept Check

1. True

3. False; On a number line, any number to the right of another number is larger than that other number.

Practice

5. 2.57

7. 6.028

9. a. 5

b. 2

c. 2, 5, 2

d. 3.0065

Applications

11. Two and eight-hundred twenty-five ten-thou-sandths

Writing & Thinking

13. Moving left to right, compare digits with the same place value. When one compared digit is larger, the corresponding number is larger.

Chapter 11.R: The Arts

11.R.1 Exercises

Concept Check

1. True

3. False; In some cases, fractions can be converted to decimal form without losing accuracy.

Practice

5. $\dfrac{9}{50}$

7. 6.67

9. 0.7

Applications

11. 17.92 inches

Writing & Thinking

13. For the numerator, write the whole number formed by all the digits of the decimal number, and for the denominator, write the power of 10 that corresponds to the rightmost digit. Reduce the fraction, if possible.

11.R.2 Exercises

1. True

3. False; The ratio 8:2 can be reduced to the ratio 4:1.

5. False; A proportion is a statement that two ratios are equal.

7. True

9. True

Practice

11. $\dfrac{1}{3}$

13. 60 miles per hour

15. True

17. $x = 10$

Applications

19. $\dfrac{3 \text{ fat grams}}{10 \text{ grams}}$

21. They are the same.

Writing & Thinking

23. The ratio can be written as $\dfrac{5}{3}$, 5:3, or 5 to 3. The preferred method is as a fraction because it is easier to simplify and manipulate in mathematical operations.

25. A proportion has been set up correctly if the same units are in the same location in both ratios.

11.R.3 Exercises

Concept Check

1. True

3. True

5. True

Practice

7. **a.** Straight

   b. Right

   c. Acute

   d. Obtuse

9. **a.** 150°

   b. Yes; $\angle 2$ and $\angle 3$ are supplementary.

   c. $\angle 1$ and $\angle 3$; $\angle 2$ and $\angle 4$

   d. $\angle 1$ and $\angle 2$; $\angle 2$ and $\angle 3$; $\angle 3$ and $\angle 4$; $\angle 1$ and $\angle 4$

11. Equilateral

Applications

13. **a.** $m \angle Z = 80°$

    b. Acute

    c. $\overline{YZ}$

    d. $\overline{XZ}$ and $\overline{XY}$

    e. No, no angle is 90°

11.R.4 Exercises

Concept Check

1. False; If there is no exponent written, the exponent is assumed to be 1.

3. True

Practice

5. y^{11}

7. $\dfrac{1}{x}$

Application

9. 2^8 GBs

11.R.5 Exercises

Concept Check

1. True

3. False; You would need to multiply by $\sqrt[3]{a^2}$.

Practice

5. $\dfrac{-3\sqrt{7}}{7}$

7. $\sqrt{6} + 2$

Application

9. $r = \dfrac{\sqrt{V\pi h}}{\pi h}$

Writing & Thinking

11. Multiply both the numerator and the denominator by the conjugate of the denominator. This works because multiplying the denominator by its conjugate results in an expression with no square roots. Answers will vary.

11.R.6 Exercises

Concept Check

1. True

3. False; Two real solutions

Practice

5. 68; Two real solutions

7. $x = -2 \pm 2\sqrt{2}$

9. $x = 1, \dfrac{4}{3}$

Writing & Thinking

11. $x^4 - 13x^2 + 36 = 0$; multiplied $(x-2)(x+2)(x-3)(x+3)$

Chapter 12.R: Sports

12.R.1 Exercises

Concept Check

1. False; Equals 81

3. False; 7^0 is 1.

Practice

5. **a.** 2 **b.** 3 **c.** 8

7. 2

Applications

9. **a.** No. Here it shows that we are only dividing the old trading cards by 6 friends versus both the old and new trading cards by 6 friends.

   b. 522; $\dfrac{15 \cdot 10 \cdot 20 + 132}{6}$

Writing & Thinking

11. If addition is within parentheses (or other grouping symbols), addition would be performed first.

12.R.2 Exercises

1. True

3. False; The ratio 8:2 can be reduced to the ratio 4:1.

5. False; A proportion is a statement that two ratios are equal.

7. True

9. True

Practice

11. $\dfrac{1}{3}$

13. 60 miles per hour

15. True

17. $x = 10$

Applications

19. $\dfrac{3 \text{ fat grams}}{10 \text{ grams}}$

21. They are the same.

Writing & Thinking

23. The ratio can be written as $\frac{5}{3}$, 5:3, or 5 to 3. The preferred method is as a fraction because it is easier to simplify and manipulate in mathematical operations.

25. A proportion has been set up correctly if the same units are in the same location in both ratios.

12.R.3 Exercises

Concept Check

1. True

3. False; In the term "12a," 12 is the coefficient.

Practice

5. -5, 3, and 8 are like terms; $7x$ and $9x$ are like terms.

7. $10x$

9. $3x + 4$; 13

Applications

11. $50,000

Writing & Thinking

13. Like terms have the same variables with the same exponents. For example, $4a^2bc^3$ and $-3a^2bc^3$ are like terms. Unlike terms either have different variables or possibly the same variables with different exponents. For example $6ab$ and $-9a^2b$ are unlike terms and $5xy$ and $13ax$ are unlike terms.

12.R.4 Exercises

Concept Check

1. True

3. True

Practice

5. 8

7. 13

Applications

9. $93.22

Writing & Thinking

11. Colby would need to know that there are 3 feet in a yard and 5280 feet in a mile.

12.R.5 Exercises

Concept Check

1. False; To change from smaller units to larger units, division must be used.

3. False; In metric units, a square that is 1 centimeter long on each side is said to have an area of 1 square centimeter.

Practice

5. 0.01977

7. 1300;130 000

Applications

9. 1750 railroad ties

Writing & Thinking

11. Each category of metric units has a base unit. The prefixes determine how many or what fraction of the base unit is being used. For example, the basic unit of length is meter and a millimeter is 1/1000 of a meter, a centimeter is 1/100 of a meter, and a kilometer is 1000 meters.

12.R.6 Exercises

Concept Check

1. False; Water freezes at 32 degrees Fahrenheit.

3. False; A 5k (km) run is shorter than a 5 mile run.

Practice

5. 77

7. 19.35

9. 72.75

Applications

11. 226.3 km

Chapter 13.R: Graph Theory

13.R.1 Exercises

Concept Check

1. False; The addition and multiplication principles of equality can be used with decimal or fractional coefficients.

3. True

Practice

5. $x = -3$

7. $x = -\dfrac{27}{10}$

Applications

9. 14,000 tickets per hour

Writing & Thinking

11. a. The 4 should have been multiplied by 3 so that the 3 was distributed over the entire left-hand side of the equation; Correct answer is $x = 15$.

b. 3 should be subtracted from each side, not from each term, and $5x - 3$ doesn't simplify to $2x$; Correct answer is $x = \dfrac{8}{5}$.

13.R.2 Exercises

Concept Check

1. True

3. True

5.

7. 0, 4, 8

9. True

Applications

11. -4500 meters

Writing & Thinking

13. If y is a negative number then $-y$ represents a positive number. For example, if $y = -2$, then $-y = -(-2) = 2$.

Chapter 14.R: Number Theory

14.R.1 Exercises

Concept Check

1. False; The numbers being multiplied are called factors.

3. True

Practice

5. 2352

7. Associative property of multiplication

9. 40 square centimeters

Applications

11. 8928 slices of bread

Writing & Thinking

13. Because any number multiplied by 1 results in the original number.

14.R.2 Exercises

Concept Check

1. False; If a division problem has a zero remainder...

3. False; $12 \div 0$ is undefined.

Practice

5. 0

7. 9

Applications

9. 16 grams

Writing & Thinking

11. To check a division problem, multiply the quotient and divisor, and then add the remainder. The result should equal the original dividend.

14.R.3 Exercises

Concept Check

1. True

3. False; 7605 is divisible by 5.

Practice

5. 3, 5

7. None

Applications

9. 5 people would raise $2480 each; 10 people would raise $1240 each.

Writing & Thinking

11. **a.** 30, 45; Answers will vary.

    b. 9, 12; Answers will vary.

    c. 10, 25; Answers will vary.

14.R.4 Exercises

Concept Check

1. False; If there is no exponent written, the exponent is assumed to be 1.

3. True

Practice

5. y^{11}

7. $\dfrac{1}{x}$

Application

9. 2^8 GBs

14.R.5 Exercises

Concept Check

1. True

3. False; The rules for exponents can be applied in any order, resulting in the same answer.

Practice

5. 64

7. $-\dfrac{2y^6}{27x^{15}}$

9. $\dfrac{y^8}{16x^8}$

14.R.6 Exercises

Concept Check

1. False; If the original number is negative, the principal square root will not be the same as the original number.

3. False; The radicand is underneath the radical symbol.

Practice

5. 7

7. 10

9. 0.2

Application

11. **a.** 4 cm

    b. 20 cm

Writing & Thinking

13. Cubing a negative real number is equivalent to multiplying a negative number by itself 3 times. The product of three negative numbers is negative.

Notes

Notes

Notes

Notes

Notes

Notes